FIREFIGHTERS FROM THE HEART

True Stories and Lessons Learned

Steve Chikerotis

THOMSON

DELMAR LEARNING

Australia • Brazil • Canada • Mexico • Singapore • Spain • United Kingdom • United States

THOMSON

DELMAR LEARNING

Firefighters from the Heart: True Stories and Lessons Learned

Steve Chikerotis

Vice President, Technology and Trades ABU: David Garza

Director of Learning Solutions: Sandy Clark

Acquisitions Editor: Alison Weintraub

Product Manager: Jennifer A. Thompson

Marketing Director: Debbie Yarnell

Channel Manager: Erin Coffin

Marketing Specialist: Penelope Crosby

Production Editor: Toni Hansen

Editorial Assistant: Maria Conto

Library of Congress Cataloging-in-Publication Data
Chikerotis, Steve.
Firefighters from the heart : true stories and lessons learned / Steve Chikerotis.
  p. cm.
Includes bibliographical references and index.
ISBN 1-4180-1423-0
(alk. paper)
1. Fire extinction—Case studies. 2. Fire fighters—Case studies. 3. Lifesaving at fires—Case studies. I. Title.

TH9310.5.C45 2006
628.9'25—dc22
                2006006957

Dedication

This book is dedicated to all of my brothers and sisters in the Fire Service. Throughout this great nation, firefighters stand ready to do battle on a second's notice. These warriors are always ready to answer that bell even though they never know what awaits them at the end of the line. Firefighters and Fire Departments come in all shapes and sizes. Each is different and unique, yet each is similar and shares a common bond. That they are willing to risk their lives for total strangers daily makes firefighters a truly noble breed.

Most of the acts of bravery are concealed from the public eye by the black smoke in which they work, but the citizens know firefighters are there when needed. Let us never forget the firefighters who came before and showed us the way, and especially those who made the supreme sacrifice.

I am truly proud to call these firefighters my brothers and sisters. Stay safe, my friends. This book is dedicated to you.

CONTENTS

PREFACE

Traffic parts like the Red Sea, and all heads turn to the sounds of air horns. Soon flashing lights can be seen reflecting off storefront windows. Here they come in all their glory—proud firefighters in a parade of shiny apparatus. Sirens scream as the big trucks seem to dance through the traffic. In the distance, a billowy cloud of black smoke marks their destination. Children wave and watch in awe as their heroes pass by. Some dream, "That could be me some day." This is our story, *Firefighters from the Heart.*

Welcome, friends. My name is Steve Chikerotis. I am a 27-year member of the Chicago Fire Department and am currently a Battalion Chief. This book contains 46 of the most exciting stories you will ever read. Each story details an actual event, and photographs accompany several of them. I wrote some of them from my own experiences, and the rest after spending well over a hundred hours interviewing some of the brave men and women in the fire service.

This book will take you into the zero visibility of dark smoke and allow you to feel the heart-pounding excitement that firefighters experience as they slip into the darkness to battle the unknown. To firefighters the stories are designed to serve as an educational tool. The lessons learned in each story can save a life, maybe your own. To the average citizen this will be a real world lesson on what firefighters do for a living. You will come away with a true understanding of what it takes to be a firefighter.

Each story comes from the firefighter's heart; as they bare their souls and share their emotions with you. The stories come from across the nation, from cities and towns, big and small. The stories have shaped and molded the careers of each firefighter. Some of the incidents happened yesterday, and some several years ago, but all are relevant today.

Firefighting is one of the most exciting occupations on earth. Risking their lives to save others is what firefighters do and have been doing for more than one hundred years in the United States. Running into burning buildings while others are running out was not invented in New York City on September 11, 2001. The heroic action of the 343 firefighters who gave their lives in the line of duty on that tragic day drew worldwide attention to what firefighters do every day.

In this book you will ride along with us on every fire, smell the smoke, and feel the heat. You will feel the pride, the pain, and the passion, as this group of firefighters takes you into a world seldom seen. Mask up, strap on your helmet and enjoy reading *Firefighters from the Heart.*

ACKNOWLEDGMENTS

I would like to acknowledge the many people who have made this book possible. First, I am indebted to all of my brothers and sisters who shared their stories with me. These firefighters bared their souls and spoke from the heart. I had known some of them for years, and I met others through the interviews. All treated me like the family that firefighters are.

Because of the selfless action of this group, lives will continue to be saved over the years. Readers of this book will see firefighters in a whole new light and get a glimpse into their world.

I would like to thank my contributing photographers who shared the visual images that accompanied the stories. To Billy Noonan from the Boston Fire Department, Lieutenant Randy Clay and photographer Bill Burnham of the Chicago Fire Department, Dave Berger, Mike Prendergast, Walter Mitchell, and John Genova, thank you, guys. Your talents and exciting pictures brought my project to life. I especially thank my good friend Jim Regan for all of his help and support. Jim provided several photos and also contact numbers of some of the firefighters. One of these numbers led me to Joe Higgins of the Fire Department of New York Communications Center. Joe and his wife welcomed Jim and me into his home and gave me a grand tour of the Fire Department of New York as I researched my project. This trip included a few emotional interviews and an equally emotional visit back to the hallowed ground of the World Trade Center disaster. The publisher and I would like to thank the many fire service folks who reviewed these stories as they were being developed. Their insights proved to be invaluable.

Mike Cox
Instructor
Florida State Fire College
Ocala, FL

Pat McAuliff
Director of Fire Science
Collin County Community College
McKinney, TX

Pete Evers
Captain
Auburn City Fire Department
Auburn, CA

M. B. Oliver
Fire Technology Director
Midland College
Midland, TX

Gail Ownby-Hughes
University of Alaska
Anchorage, AK

Jeff Simpkins
Director Fire Service Extension
West Virginia University Fire
Service Extension
Morgantown, WV

Clint Smoke
Fire Science Program Chair
(emeritus)
Asheville-Buncombe Technical
Community College
Asheville, NC

To all of the people at Thomson Delmar Learning, especially Jennifer Thompson, Toni Hansen, Alison Weintraub, Stacey Wiktorek, and Jennifer Douglas as well as Mou Sen Gupta at Interactive Composition Corporation for believing in me every step of the way.

Last, I would like to thank my family—my wife Mary and four sons Steve, Luke, Dave, and Pete—and also Steve's wife Lisa and Luke's future wife Nancy. And the support of my friends has been invaluable. I would like to give a special thanks to my mother, Edna Miller, for her constant motivation. Mom was diagnosed with lung cancer as I started this project and, though she was undergoing treatment, I never once heard her complain. Every day she would admonish me to keep writing. I'm happy to say she fought hard and is winning her battle with cancer. Mom, I'm proud of you. Thanks for being a constant inspiration.

INTRODUCTION

Excitement and danger seem to go hand in hand, and the life of a fire-fighter sometimes has a cost. Each year an average of more than one hundred firefighters die in the line of duty in the United States. After each death, the doorbell rings and a solemn-faced fire chief or chaplain is on the other side. These people have the grim task of telling family members that their loved one is never coming home. An emphasis of this book is on safety, and measures that can increase safety in the Fire Service.

Those of us in the Fire Service face a great challenge, and today is the most dangerous time in history to be a firefighter. I base this belief on three reasons: Fires burn hotter, buildings collapse more quickly, and firefighters respond to fewer fires. This statement may be confusing, so let me explain.

First, fires do burn hotter than ever because more plastics and synthetics are introduced each year. These plastics burn much hotter and faster than ordinary combustibles such as hard wood. The Fire Services uses the standard time–temperature curve to represent actual fire temperatures in relation to time. This curve, however, was designed several years ago when the use of plastics was still in its infancy. Therefore, we must expect actual fire temperatures to far exceed the 1000 degrees in five minutes as given in the curve. The danger stems from the fact that the steel commonly used in building construction fails at 1000 degrees or less.

Prior to the 1980s, most structural steel was massive. The rate at which steel absorbs heat and subsequently fails is directly related to its mass. Today, most steel used in buildings is lightweight—such as in metal gusset plates, which are as thin as a postcard. Lightweight is not a good thing in the structural components of a building on fire because it absorbs heat rapidly and fails. Lightweight construction is cost-effective, but it is not life-effective.

Lightweight truss construction has become common in both residential and commercial buildings. Many of these buildings are ready to collapse within five minutes of structural involvement. This should give a family with working smoke detectors time to flee, and they often are leaving about the time firefighters are arriving. If they are not, a primary search is a priority for firefighters.

Firefighters are responding to fewer fires than in past years. This is good from the citizens' perspective, but it does require more effort for firefighters to be prepared for battle when fire calls are infrequent. Inactivity can dull our skills and instincts, so drills and training become more important in our daily routine. The avenues we have to combat the inherent dangers of the job are protective clothing, tools, equipment—and especially education. We receive education formally in structured classes and less formally in training and drilling in our firehouses and by reading articles, or books.

The most valuable education firefighters receive comes from the University of the Hallways. The more hallways a firefighter crawls, the sharper his or her knowledge base and skills become. We develop instincts and are more capable of making the split-second decisions. We must be able to think on our feet and adapt quickly to the ever-changing conditions on the fireground.

Firefighters have been building a knowledge base for years through on-the-job experiences. Many of us have benefited from the mentorship of firefighters, officers, and chiefs early in our careers. Over the years, we learn lessons from every close call we have, and every injury we suffered or witnessed.

Over the years, experienced firefighters have shared their stories with their peers, and the knowledge packed into these stories has saved lives. This is the basic premise behind the study of risk management—a prominent theme in this book. The basic concept of risk management is to study past history so we can better predict problems that may arise in the future and prevent them. A good piece of advice is: Predictable is preventable. Utilizing risk management can and does save lives.

Lessons Learned

The "heart" of this book is in the lessons the firefighters learned from their experiences. At times, these experiences challenged their perspectives and attitudes about their role as as firefighters. In this respect, the book represents an excellent training tool for incoming firefighters, as well as the veteran firefighters who teach them. Within these pages is a story for everyone. The following features conclude each story to encourage reflection specifically in the classroom environment.

- A bulleted list of *Lessons Learned* at the end of each story, highlighting the main teaching points the firefighters wish to communicate through their stories.

- Specific *Discussion Questions,* which provide an opportunity for incoming firefighters to discuss the specific story and develop critical-thinking and problem-solving skills for realities they will face on the job.

In this book, the firefighters share more than their stories. They provide rare insights into their very souls, their personal sacrifices, their struggles, their character development. After reading the book and reflecting upon the stories, incoming firefighters—and anyone who wants to better understand the challenges and day-to-day life of firefighters—will emerge more informed and appreciative of this often unheralded occupation.

Learning the Lingo

As you read through these stories, either as a fan of firefighting or as an aspiring firefighter, you may come across some terms that are unfamiliar to you. To help you become better engaged with the firefighters and their stories, we have provided you with a list of firefighting terms and their definitions. We hope this leads you down a path of better understanding of the fire service:

backdraft   A backdraft is commonly referred to as a smoke explosion. More clearly, it is an explosion caused by oxygen suddenly being admitted to a confined area that is superheated and contains a large volume of combustible vapors and smoke.

bowstring truss   The bowstring is a type of truss used to support roofs and is recognizable by the distinctive arched shape. When a large volume of fire contacts these trusses, a large-area collapse is to be expected.

CISD (Critical Incident Stress Debriefing) / CISM (Critical Incident Stress Management)   CISD is a step in the overall CISM. The fire service has recognized that stress reactions resulting from high-stress emergency situations can have a negative impact on firefighters' work and life. CISD teams typically are made up of trained fire department members and guided by mental health care professionals.

cockloft   The cockloft is the area between the top floor ceiling and the roof deck in flat-roof buildings.

copping stones   Copping stones are tiles made of stone or clay that top a masonry wall to protect it from the elements.

engine companies   An engine company is a group of firefighters that responds to incidents in a pumper, or engine (the apparatus that carries the hose). At structural fires their main duties are rescue and extinguishment. Engine companies rescue occupants by confining the fire and protecting unburned portions of the building and stairways.

fireground   The fireground is the working area surrounding a burning structure or incident.

fireground channel   The fireground channel is the radio channel that firefighters use when working at an incident.

flashover   Flashovers are common occurrences at structural fires, in which the temperature within a room or space is heated to a point at which all of the combustible objects within that space ignite almost simultaneously. Flashovers happen quickly, and firefighters in full protective gear can survive only a few seconds in these conditions.

halligan bar   The halligan bar is a common pry bar that firefighters use to force entry.

lightweight construction   Lightweight construction is commonly used in modern buildings. Because it can collapse in the very early stages of a fire, it is dangerous to responding firefighters.

master-stream device   This firefighting device flows more than 350 gallons per minute and is used on defensive attacks from the outside of the building. Use of these devices is common when there is a strong probability of building collapse.

overhaul   Overhaul is the act of finding and exposing any hidden pockets of fire or smoldering material to ensure that the fire is completely extinguished.

pike pole   The pike pole has a hook at the end for opening ceilings and sidewalls to expose the hidden fire. It commonly is found in lengths of 6 to 16 feet.

rescue squad   The rescue squad is a company of firefighters who respond in an apparatus with a large assortment of rescue equipment. As the name implies, the primary duty is rescue. The most common rescue performed by these companies is fire rescue. They also are equipped and trained for extrication from vehicles and machinery, high-angle rope rescue, underwater SCUBA rescue, and rescue from trench collapse, confined space, hazardous materials, and the like.

RIT teams   RIT stands for Rapid Intervention Teams, fire companies assigned to stand by at incidents equipped and ready to go when a Mayday is called in response to trapped or missing firefighters. Their sole purpose is firefighter rescue.

rollover   Rollover is the term used to describe burning fire gases in the upper regions of a room or structure. If left uncontrolled, this condition will lead to flashover.

sectors   The four sides of a building are divided into sectors, with sector 1 or (A) being the front of the building. Moving clockwise sector 2 or (B) is the left side, sector 3 or (C) is the rear, and sector 4 or (D) is the right side.

SCBA   SCBA stands for Self-Contained Breathing Apparatus.

taxpayer   "Taxpayer" is firefighter slang term commonly used to describe multi-tenant storefronts within the same building. Quite often, they have one or two floors of apartments above them.

triage   Triage is the sorting of two or more victims based on the severity of their condition to establish treatment priorities.

truck company   A truck company consists of firefighters who respond on an aerial ladder apparatus, which primarily carries an assortment of ladders. The main duties of truck companies are rescue, ventilation, and forcible entry.

∽ Praise for *Firefighters from the Heart* ∽

"The author did an excellent job in addressing some of the major issues that involve firefighters today . . . the topics covered in the book are real world experiences and provide an excellent illustration of life in the fire service."
—Captain Pete Evers, Auburn City Fire Department, CA

"I'm excited to see a book that ties real life experiences to firefighter training. What I liked most about the book was the balance between entertainment and training. The author has done a good job of selecting stories that are fun and exciting to read, yet manage to teach a lesson for those who are willing to learn. . . . I truly believe his words are from the heart."
—Pat McAuliff, Director of Fire Science, Collin County Community College, TX

"These stories apply themselves to administration, training, and fire-fighting points. I can use this in just about any [firefighting] course. . . . There is not one story that won't apply. Learning from these firefighters' experience is a great teaching tool."
—M. B. Oliver, Director of Fire Science Technology, Midland College, TX

"I commend this author for an enormous undertaking. . . . There is great value in the book. It will be popular among the general public as an interesting read . . . and certainly this book should be read by firefighters and fire officers. Both of these audiences will find the stories and lessons in this work interesting, informative, and entertaining.
—Clinton Smoke, Fire Science Program Chair (emeritus), Ashville-Buncombe Technical Community College, NC, Author of *Company Officer,* 2nd Edition

A Lot to Learn

Battalion Chief Steve
Chikerotis 1979
27 Years in the Fire Service

It was Super Bowl Sunday in January of 1981. I arrived at 6:30 a.m. to work a 24-hour shift on Truck 42, a busy truck company. This wasn't my usual assignment, and I was sent there only for the day. I was in my early twenties, with less than two years on the job. I had a lot to learn.

Of the ten firefighters who reported for duty that day, the only one I knew was Lieutenant Benny of the engine company, who was also there just for the day. An African American who appeared to be in his early 40s, he had a thick neck and a barrel chest—plus a great personality and a good sense of humor. He also had a reputation as a good firefighter and as an excellent officer.

A Typical Day

The day started like any other day. We got our tools and equipment ready, did the housework, cleaned our truck, and had a short company drill. We spent the rest of the morning on alert for the alarm dispatcher to call while watching all the Super Bowl hype on television. We responded to an occasional alarm but nothing of any consequence. As the day went on, I became more and more comfortable with my new friends. By the opening kickoff of the Super Bowl, every firefighter in the house was in the dayroom seated around the television.

This was my chance to strike! I pulled up a chair about two feet from the television, blocking everyone's view. I then performed the ultimate sin and changed the channel to a dance show. I sat down and pretended to be watching it.

Total silence fell over the room. Finally one of the firefighters shouted, "What do you think your doing?" Angry voices from around the room joined in.

I looked back at them. "Were you guys watching something?" I asked, adding, "Sorry guys, but I really have to watch this show. It's where I get all my good dance moves." Then I stood up and started dancing with as much rhythm as a gorilla in a phone booth.

Finally realizing the joke, everyone erupted in laughter.

Time to Go to Work

We all sat back and were enjoying the game when the alarm came in: "Engine 82 and Truck 42, take in. . . ." The address on South Cottage Grove was only a few blocks from the firehouse. As soon as we rolled out the doors, we could smell the smoke.

"We got a hit"—a working fire—several of us yelled at once. When the engine and truck companies turned the corner, we could see the monster that awaited us.

As we pulled up to the fire building, I could hear Benny's voice on the radio: "Engine 82 and Truck 42 are on the scene. We have a one-story commercial, bowstring truss construction, 75 by 125. We've got a fire. We're northbound."

Heavy, dark smoke was pouring out of the rear of the large building.

Heavy, dark smoke was pouring out of the rear of the large building. It was a supermarket that had been converted some years back to a federal government building.

My officer yelled back at my partner and me, "Billy, you and Steve get that rear door."

I grabbed a K-12 saw and headed toward the rear, Billy had a halligan bar and a sledge hammer. I started cutting on the well fortified rear door. The smoke was pumping hard through cracks around the door, and the rear of the building seemed to disappear in the thick black smoke.

My heart racing faster than the engine of the saw, I said to myself, "It doesn't get much better than this!" After cutting through three heavy steel hinges and two padlocks, I shut down the saw and helped Billy force open the burglar gates, then the rear door. After several minutes of exhausting work, we yelled to the lieutenant's engine company, "We're in!"

The engine company and its huge 2½-inch hoseline seemed to get swallowed up by the smoke. Despite the fact that it was a chilly January night, we were drenched in sweat.

"Now it's time to really go to work," I told myself, as Billy and I followed the engine company inside. The smoke was so thick that the visibility inside was zero, but, surprisingly, it was not hot at all.

Billy and I were using 12-foot-long pike poles to pull down the false ceiling tiles. A sea of fire could be seen above us in the wooden bowstring trusses. The firefighter handling the nozzle behind us was

> *The smoke was so thick that the visibility inside was zero.*

throwing 300 gallons per minute at the fire, but that only seemed to make the fire angrier.

"This sure beats the Super Bowl!" I yelled to my partner.

Time to Retreat

"Back this line out!" the lieutenant shouted. His education and instincts honed over a twenty-year career told him it was time to retreat.

"Oh no," I thought, "we're not going to give up are we?" After all, I had almost a whole three years on the job. I grabbed the lieutenant's arm and said, "Hey lieutenant, I think we're winning."

The lieutenant repeated in no uncertain terms, "I didn't ask your opinion, now help us back the line out!"

I did exactly as he said.

Once out of the building, we set up a portable deluge gun—a water cannon—in the alley off the corner of the building. Now we were taking a defensive stance, giving up the fire building and protecting the neighboring buildings. I grimaced through the smoke. I was sure that we could have extinguished the fire if we were given a few more minutes.

The Lesson

Without warning—"Booomm"—the roof collapsed over the entire rear of the building. The area where we had been working only minutes before was now a pile of debris amidst flames shooting 50 feet into the air. The fire had come to us.

I looked over at the lieutenant. "Thank you," I said humbly.

He just smiled and nodded.

Back at the firehouse, I thanked him again and apologized for questioning his judgment.

"Steve, you're a good, tough fireman, but you have a lot to learn," he advised. "Learn the job and you'll do the same thing for other firefighters someday."

Six firefighter lives were saved that day.

Six firefighter lives were saved that day, not by a dramatic rescue but by a fine officer who was a student of the game. That night made me realize that education and using your head is more important than physical toughness. On that day I also became a student of the game.

Lessons Learned

- **A false sense of security poses dangers.** The lack of heat inside the building gave us a false sense of security. We found out that the large half-barrel shape of the roof acted as a large vessel collecting the heat and smoke. Firefighters several feet below at floor level often face light smoke conditions and very little heat. This can easily cause firefighters to mistake the severity of the situation.

- **Large area collapse.** The trusses above us spanned 75 feet and were spaced 20 feet on center. Therefore, the failure of only one truss would have caused the collapse of a section of roof 40 feet by 75 feet.

- **Sometimes there is no warning sign.** When trusses are involved in a fire, they will collapse, and usually without any warning. Eight years after the above incident, five firefighters died as a result of bowstring truss collapse in Hackensack, New Jersey.

- **Front and rear walls pose a special danger.** The front and rear walls of bowstring trusses are the most vulnerable to a 90-degree collapse. These are always dangerous areas.

- **Observe collapse zones.** Once firefighters are backed out to a defensive position, we must observe collapse zones of 1½ times the height of the walls. Stay out of alleys and gangways.

- **We need to be students of the game.** This means continuing to learn until retirement.

- **Safety is every firefighter's responsibility.** Everyone has a role in the safety of the team; if you see something unsafe report it to others. Officers, if you feel your firefighters are in danger summon the courage to move them. Err on the side of safety.

- **Utilize Risk Management.** Significant risk should only be taken when there is the potential to save endangered lives. Reduce the risk when only property can be saved and take no risk when there is no possibility to save lives or property.

Discussion Questions

1. What triggered the lieutenant's order for the firefighters to back out of the building?

2. What caused the conditions to be much worse than they seemed to the interior crew?

3. What are the most dangerous walls in bowstring truss collapse? Why?

Miracle Rescue: The Victim

Firefighter Bob Dodovitch,
Rescue Squad 2
17 Years in the Fire Service

"**R**ing, Ring, Ring." The house bell pierced the silence in the bunkroom at 1:30 a.m., just as I was starting to fall asleep.

"Thank goodness," I thought, as I pulled the blankets over my head. "It's for the engine company, not me."

Our Rescue Squad had already been to four fires during this shift, and I was still wet and dirty from the last one. The overhead door opened for the engine company, and the air filled the bunkroom. Feeling the night air and smelling the diesel fumes I settled deeper into the blankets, muttering, "Better them than us."

A few minutes later the speaker crackled, "Squad 2, start into the still alarm fire at 2814 Milwaukee . . ." (first alarm), and the bunkroom came to life.

Off and Running

"Ring, Ring, Ring, Ring." Four bells were ringing as we ran to the squad. The adrenalin rush quickly brought our tired crew to life. We took great pride in our quick pushouts, and within seconds the squad, with its crew of six, was headed towards the fire. The engine company (one engine and one truck) had responded to an automatic alarm, found heavy fire conditions in the back of a currency exchange building, and escalated the incident to a full still alarm (two engines, two trucks, squad, and battalion chief). Firefighters in the first truck had easily gained entry into the front lobby, but the fire was in the secure back area of the currency exchange.

This was how the fire looked when we arrived. Note the 20-foot straight-frame ladder being taken to the rear of the building to give the team a second way off the roof. This ladder would soon be used to save my life.

Our squad stopped in front of the building, and we observed the fire conditions as we jumped out. The currency exchange was closed—as could be expected at this late hour. The building was 25 feet × 100 feet and appeared to be of ordinary construction (masonry walls and wood joist). We could see heavy black smoke exiting in small puffs from the tightly locked building, and the glow of fire through the bulletproof customer service windows.

My officer instructed me to give Truck 13 a hand in venting the roof. I had spent nine years with this same truck crew before transferring to Squad 2 four years ago.

Grabbing an axe, a pike pole, and a K-12 saw, I quickly climbed the truck's aerial ladder to the roof. The rest of the squad split into two teams and attempted forcible entry through the fortified front and back doors. This challenge was like breaking into a bank vault, and it obviously was going to take some time. Two squad members used our Arcair® torch on the front vault-type door. The temperature of this torch reaches 6700 degrees F. at the tip and it will cut almost anything. This time, though, it appeared to meet its match; sparks flew but the door remained secure.

Our officer and two firefighters were having similar bad fortune with the back door. The blade had little effect. My officer called for an additional saw, and I lent him mine by hooking the shoulder strap with the pike pole. Members of Truck 13 were already venting the roof with their saw, and I started to assist them with my axe.

To reach the fire, firefighters attempted to cut through the door that led to the secure area of the currency exchange.

"Squad 2 to Squad 2 roof. All the fire is in the rear. Open it over the back," my lieutenant commanded on the radio.

"Squad 2 roof, message received," I replied and moved with my partner Greg from Truck 13 toward the back of the roof.

To get to the rear of the roof, I had to step up onto and over an I-beam that supported an air-conditioning unit. From this elevated vantage point, I happened to look across Milwaukee Avenue. On a wall directly across the street was a mural of three firefighters from Truck 58 who had died in the line of duty about fifteen years ago. The memorial marks the location where these three firefighters died when a roof they were attempting to ventilate collapsed during a fire at an electronics store.

I've seen this memorial many times, but this time a tingle went through my body. Maybe it was because Truck 58, the same company, had just thrown a 20-foot straight ladder to the rear to give us a second way off the roof. We moved to the back, and soon the saw motor was screaming and chunks of gravel and tar shot behind the saw like machinegun fire. Two other truck members and I used our axes to cut away on the thick asphalt roof. The smoke was getting thicker with each swing of the axe.

> *The smoke was getting thicker with each swing of the axe.*

About this time, another truck company appeared on the adjoining roof. The automatic alarm from the store next door had gone off and sounded for Truck 35. This could have, and should have, been reported as an alarm tripped by our fire, and the firefighters would have returned to their nice warm beds. After all, it was now close to 2:00 a.m. and that crew had been busy all day, too. Still, they decided to stay and help. An 8-foot cyclone fence stood between their roof and ours, but it had a slit through which we could pass.

After a short while, Keith the officer of Truck 35 said, "Let's back off this roof—it feels a little spongy."

Two members of Truck 13 listened and immediately headed toward the ladder. I stomped my foot and it felt solid, so my partner and I continued to swing our axes.

"We'll be off in a minute, we just want to finish this hole." We said this not out of disrespect. It was just that the roof felt solid, and we needed only another minute to finish.

About a minute later the battalion chief came on the radio and ordered us off the roof.

"He must see something we don't see," I said to my partner. "Let's go."

Following my partner, we moved toward the slit in the fence. There we could step over a parapet wall and onto another section of roof where Truck 58 had placed the 20-foot ladder. My partner was about to step through the fence and I was two steps behind him when I raised the microphone to my mouth and announced, "Squad 2 to Battalion 7—everyone is off the roof." I took one more step.

No Way Out

Crash! The roof gave way as if someone had pulled a rug out from under me. My partner clung to the parapet wall by one arm, and I rode the roof down into the inferno. Everything was happening in slow motion. Thick smoke swallowed me up as I slid down the large section of roof and somehow landed on my feet at the bottom. Now I was locked in the fortified rear section of the currency exchange, and I knew the firefighters

> *Everything was happening in slow motion.*

hadn't yet gained entry to this area. Everything was immersed in total blackness, and the heat was intolerable.

A large section of the collapsed roof was still hanging straight down. For the moment, at least, it was shielding me from the intense

Within seconds the flames would engulf me.

fire on the other side. I tried to move, but there was no way out. Within seconds the flames would engulf me. I was going to die!

My officer had taught me never to panic, never to quit—to slow down and think. Immediately I grabbed my radio. "Emergency! Emergency!" I yelled up into the blackness. "Greg, get me out of here!"

Trapped in the tight space, I could move only a few feet in either direction. In the midst of total silence, I continued to yell, "Help! Get me out of here!"

In the total blackness I tried to move and I bumped into a wall. For a second I stood there with my face to the wall, and I could clearly see my family—my wife and my three kids. I was standing face-to-face with my little guy, who was only one year old.

The heat was getting more intense by the second.

"This can't be happening to me! I can't die here!" Now the flames were burning through the collapsed section that had been shielding me. The heat was getting more intense by the second. I kept moving, looking for a way out. I had to fight. I had to get back home to my family.

Through the Darkness an Angel Calls

Suddenly from above I heard, "Bobby, move to your right." I moved to my right, and something hit me in the head.

Just then the whole area lit up in flames and I saw it, the most beautiful sight I had ever seen—a ladder. Half of the ladder was engulfed in the flames as I dove for it. I don't remember climbing the ladder, just the feeling that I was going to make it. At this point I didn't care if I ended up in a burn unit, because I knew I was going to live. I knew I was going to see my family again.

*As I reached the top of the ladder, firefighters of Truck 35
pulled me out of the hole and threw me onto the
snow-covered roof to extinguish the fire.*

With my eyes and mouth shut tightly to protect them from the flames,
I raced up the ladder. By the time I reached the top, I was on fire. Two
firefighters from Truck 35 grabbed me, pulled me through the hole in the
fence, and slammed me onto the snow-covered roof. As they hit me with
their palms, slapping out the flames, I felt no pain. My partner Greg was
on his knees looking into my eyes.

We spoke no words, but we knew what the other was thinking: "We're
alive! We made it!"

Others were yelling, "Get an ambulance!" "We got him!" "Get a ladder." I remember climbing down a ladder and being rushed through the
snow to a waiting ambulance. The burns meant nothing to me. I'm going
home!

As it turned out, I was very lucky, receiving only small first- and
second-degree burns to my face and legs. I was treated and released from
the hospital in only a couple of hours.

To say it was an emotional homecoming is an understatement. I
hugged my family like I have never hugged them before. I didn't want to
ever let go. I know how close I came to leaving them.

An amateur photographer captured my rescue on videotape, and
within a few hours I received a copy. When I saw the video for the first
time, I realized just how much I had cheated death. Flames were spewing
out of the hole like a blowtorch when I came up that ladder totally

As firefighters guided me down the ladder, I knew I was going home!

engulfed. For the first time I could see who my rescuers were. The firefighters of Truck 35 were the ones who saved my life. They weren't even supposed to be there, but their lieutenant kept them there in case we needed help. Firefighter Marty Nolan was the one who heard my screams and, with flames in his face, somehow managed to see me. He and the others grabbed that ladder and fished me out of hell. I owe everything to these brave heroes.

For a long time I had occasional nightmares in which I saw myself shriveling up as the flames engulfed me, and I would wake up in a cold sweat. For the most part, now, my life has moved on, but the little things we take for granted as we rush through life are more important to me. My priorities are clearer, and nothing means more to me than my family. I'm still aggressive at fires, but I'm much more of a thinking firefighter.

Lessons Learned

- **Ordinary construction can collapse.** Remember the 20-minute rule: Ordinary construction (masonry and wood joist) fully involved in fire can collapse in less than 20 minutes. If we're not making progress, we must be aware of time. It's not on our side.

- **Impact load triggers collapse.** On a roof or floor that has been weakened, firefighters walking or swinging axes create impact loads that can cause a collapse.

- **Do not panic.** In bad situations, the natural reaction is to panic, but firefighters must react by staying low, thinking, and never giving up. If we keep our head, we can survive.

- **Warnings and orders are to be heeded.** If someone orders or warns us to exit a building or a rooftop, we should not question the order. If it's wrong, let it be wrong. Err on the side of safety.

- **Firefighters must think.** We must "read" the building when we arrive. At all times, we have to know where we are and where the fire is, and we have to remember that the conditions are continuously changing. Buildings on fire never get stronger. They get weaker by the second.

Discussion Questions

1. What were some of the warning signs prior to the roof collapsing?

2. This incident happened before this department had a standard Mayday procedure. Does your town have one? If so, what is it?

3. This incident happened before a Rapid Intervention Team program was implemented at this department. Could a RIT response have helped? If so, how?

Miracle Rescue: The Rescuer

Firefighter Marty Nolan,
Truck 35
23 Years in the Fire Service

Two o'clock in the morning and the bells started ringing. It seemed like we had just got to bed, we'd been busy all day—and here we go again. In a minute we were out the door on our way to an automatic alarm, nothing special, just an automatic alarm.

"Automatic alarms are usually nothing," I thought. "We'll probably be back in bed in a few minutes."

On the radio we heard that neighboring companies were working at a fire on Milwaukee Avenue. Our automatic alarm was just around the corner from that fire. As we pulled up, we could see heavy smoke behind our building. We shined our flashlights through the storefront window, and it looked clear as a bell. The smoke we saw in the rear was coming from the Milwaukee Avenue fire. The back wall of our building adjoined the building that was on fire, so our officer had us ladder the building to check the roof.

On the Roof

We stepped off the ladder onto the roof, and walked to the rear. It was now quite obvious that the fire next door had tripped our alarm accidentally. Keith, our lieutenant, relayed this information to the alarm office, and he decided to keep us on the scene for a few minutes to see if anybody needed help. The buildings butted up against each other near the rear; we were able to walk across the roof to the building on fire.

The fire was in a one-story currency exchange. A rusty old fence with barbed wire at the top surrounded the roof. The fence was collapsed in one spot and had a hole in it so my lieutenant and I were able to slip through and walk out onto the roof over the fire. Bob Dodovitch from Squad 2 and three firefighters from Truck 13 were venting the roof.

Our lieutenant thought the roof felt spongy. He ordered everyone off the roof and relayed this information to the battalion chief by radio. Our group stepped back through the hole in the fence to the safety of the neighboring roof. Two of the four firefighters followed us onto the neighboring roof. I watched as the two headed in the opposite direction towards a 20-foot, straight-frame ladder. I continued to watch as they climbed down.

One of our firefighters said to our officer, "They'd better get off there." When I turned around, two of the firefighters, Greg and Bob, were still working on the roof. Just then the chief's voice could be heard over the radio ordering everyone off the roof; finally they stopped working and started walking in our direction. We turned away to head back to our own ladder.

The Collapse

"Crash!" It happened so fast. I felt the roof shake and we all started to run, thinking it was our roof that was collapsing. Quickly we realized that the sound hadn't come from our building, and we stopped. By the time we turned around, Bob and Greg had disappeared and heavy smoke was pouring out of a huge hole. We ran back in that direction. An arm reached out of the smoke; it was Greg's, and he was fighting for his life as he clung to a wall. We grabbed his arm and pulled with all of our might, and soon up came his other arm, followed by his big body. By now, everyone in our company was sprawled on the roof, grabbing a piece of him and pulling him out. Greg crawled a few feet away into the deep snow and froze. He knew how close he had just come to dying.

"Where's Bob?" I yelled. Suddenly across the radio came the urgent plea, "Emergency! Emergency! Firefighter down!" It was Bob's voice, and his message was blood-curdling. A couple of us peered into the hole that spewed thick black smoke. We couldn't see anything. We called Bob's name, with no response.

Nothing but Blackness

"We need a ladder!" I yelled, remembering the 20-foot, straight-frame ladder. We grabbed it and slid it into the hole until it landed on something firm. I tried to climb down but made it only about four rungs. The heat

The firefighters were unable to find Bob through the heavy smoke and heat as they flailed around blindly with the 20-foot straight frame ladder.

was too intense, and the smoke was choking me. Visibility was zero. I backed out onto the roof.

We yelled Bob's name and tried to see him through the smoke, but the smoke was getting heavier and hotter by the second. Suddenly the darkness was replaced by bright flames as everything lit up like someone had just poured gasoline on the fire.

"There He Is"

In the total silence, I thought, "We lost him." My heart was pounding out of my chest. "We're so close to him, and yet there's nothing we can do." I squinted as the flames were now licking my face.

"There he is!" The flames pierced the blackness and I could see him now. He was facing the wall, moving back and forth. "We only have seconds," I thought.

> *Suddenly the darkness was replaced by bright flames as everything lit up like someone had just poured gasoline on the fire.*

Grabbing the ladder, we aimed it in his direction. He seemed to be moving the wrong way, so I yelled at the top of my lungs, "Bobby, go to your right!" I could hear him screaming, "Where are you?"

When everything lit up, Marty spotted Bob and screamed at him to find the ladder.

We moved the ladder again, and I shouted again, "Bobby, go to your right." Suddenly he dove for the ladder and started climbing like a cat. My lieutenant and I dangled our arms into the flames, feeling for him.

> ## He dove for the ladder and started climbing like a cat.

Like a miracle, Bob was now at the top of the ladder, and we grabbed him. He was on fire as we pulled him out. Instinctively, we threw him on to the snow-covered roof deck and started rolling him around and slapping at the flames with our hands.

"Bobby's as tough as nails," I thought. Still, I was worried that he wasn't going to make it. I ran to the roof edge and shouted along with the others, "Get an ambulance! Get us a ladder! We got him!"

Several firefighters flocked to the area, and Bob actually climbed down the ladder himself. With assistance, he walked to the ambulance. I felt better about his chances now, but with burns, you never know how much damage has been done. Thank God we got him out.

We stayed and worked at the fire for another hour, barely talking. We were thinking about Bob and wondering about his chances.

As a postscript, we had two more close calls that night after Bob was taken away. A lieutenant stepped through a breached wall and fell down the basement stairs. Luckily his fall was to a landing only about six feet down. When we pulled him out, he was shook up but okay. Later a group of us noticed a wall bulging and moved out of the way just before a large section of wall collapsed where we had been standing.

This fire didn't look like much, and yet there were several close calls. Bottom line—you must be alert at all times. You can never lower your guard. We were informed later that day that Bobby was going to make it, and his burns didn't appear to be serious. Now we could all feel good. I still don't know how he made it out of there. Like I said, "Bobby's as tough as nails."

Lessons Learned

- **We have to learn to think on our feet.** Firefighters never know the hand that will be dealt us. We have to be better than anyone else at thinking fast, using the equipment available, and getting the job done.

- **The building isn't getting stronger.** Most firefighters die as a result of secondary collapse. The longer we're at a fire, the more probable it is that the structure will collapse because the building is continuously getting weaker. Once any part of the structure has collapsed, secondary collapse is almost certain. Firefighters working or running to the aid of others are creating impact loads, which can cause secondary collapse.

Discussion Questions

1. What modern-day tools or equipment might have helped with this rescue?

2. What probably caused the wall to collapse after the rescue?

3. What are some of the warning signs of building collapse?

Growing with the Changing Times

Retired Boston Fire
Commissioner
Leo D. Stapleton
Boston Fire Department
39 Years in the Fire Service
Author of nine fire service
books including *Thirty
Years on the Line*

The worst thing you can ever do on this job is to believe that you know it all. This job has a way of humbling us when we get too comfortable, and the lessons learned sometimes save lives, including our own. I spent thirty-nine years in the fire service. I rose through the ranks from firefighter to Fire Commissioner, and I kept learning until the day I retired. In this story I turned a negative into a positive. This incident turned out to be a catalyst that helped push the Boston Fire Department into the forefront of high-rise firefighting.

Newly Promoted Chief

In the mid-1960s I was promoted to District Fire Chief (the same rank as Battalion Chief in many fire departments). I was assigned to the South End Back Bay District of Boston. Having served as both Fire Lieutenant and Fire Captain in this district I was familiar with the types of fires typically encountered in the course of duty. This was a busy district, and some of the best firefighters worked there.

Although the district had a sizable commercial area, along with major hospitals and schools, most of the buildings were brick-and-frame apartment houses ranging from three to six stories high. The South End buildings were mostly in poor condition and were occupied by people on welfare or without visible means of support. The Back Bay buildings, although of similar construction, were in much better condition, housing much wealthier people.

Boston Grows Tall

When I was first appointed to the job in 1951, there were only two high-rise buildings in Boston, both located downtown. By 1965, the city boasted several high-rises with several more under construction including the 52 story Prudential Tower, which was nearing completion. Today the city has hundreds of such buildings.

Fire departments have Standard Operating Procedures, called SOPs, which assign the duties of fire companies responding to ordinary fires. In the 1950s, SOPs applied to low-rise buildings but none to high-rise buildings because we had no experience with them. Unknown to me, I was about to gain some firsthand experience.

Fire in a High-Rise

One comfortable spring night, after a busy day, I was getting ready to rest for a while when an alarm sounded. We were dispatched to investigate an odor in a seven-story apartment house in the Back Bay district. Three engine companies—two ladder companies and myself, as district chief—were dispatched. Because the building exceeded 70 feet tall and was constructed after 1920, the law required a 4-inch standpipe system, with 2½-inch outlets on each floor and an exterior standpipe connection. The first-due engine company was required to connect one 2½-inch line into the inlets but usually does not fill the line unless an actual fire is in progress.

On my arrival, the first engine was already in position at the standpipe inlet, and a firefighter was connecting to the inlet. The first truck company was in front of the building preparing to raise its 100-foot aerial ladder to the roof, as required at a potential building fire if within the ladder's reach. One member would ascend to the roof unless the incident commander ordered otherwise.

This building had an elevator, unusual within the district because most apartment houses at that time were walk-ups. When I approached the main entrance, several residents were already out in front. They said that the people living on the top floor weren't home and there was an unusual smell near their apartment. I instructed the engine company to take the

2½″ donut rolls and nozzle and ascend via the stairway to the sixth floor, which was the floor below the potential fire.

Learning a New Game

Up to that point I was operating properly. Then, because no smoke was visible and the smell could have come from anything, I entered the elevator with the ladder company officer and two firefighters. At this time I made an error in judgment, and we took the elevator to the top floor. After this incident SOP's were written for firefighters investigating a fire to exit three floors below the reported floor and walk up the stairs.

When we arrived at the floor and the elevator door opened, thick black smoke slapped us to the floor. We found ourselves inside the apartment with the actual fire, a fully involved kitchen fire off to our left. Because the heat was intense and the smoke was banked down to the floor, we had to evacuate the elevator. It was too hot to remain inside. The firefighters all had service filter masks, the department standard at that time, but I had neglected to take one myself—another error in judgment.

The four of us stayed close together and followed a wall to our right. We knew we had to find a door quickly so we could use it as a protective barrier to buy us some valuable time. With our faces low to the floor, our group tried to crawl away from the heat. The ceiling was now lighting up over our heads, so we didn't have long.

One of the firefighters found a doorway. "It's a bathroom," he yelled.

"Keep going," I shouted back, deciding that being trapped in a bathroom with three other firefighters on the seventh floor wasn't where we wanted to be. We continued following the wall down the hallway while flames rolled over our heads and started to lap down. I recognized these all-too-familiar signs: Flashover would happen any minute now.

Then one of the firefighters in front of me yelled, "I found the door!" At the extreme end of the long hallway, this door led to the only stairway in the building, and we managed to tumble down to safety.

Safety in the Stairway

Now safe in the stairway, I used my portable radio to notify the engine crew of the fire conditions. I directed my aide to instruct the roof man to vent over the stairway and the elevator shaft. Within a few minutes the engine had advanced, charged the line, and managed to attack the fire successfully.

Once the roof was opened, I ordered the aerial to be retracted and dropped into the top floor. I ordered the third engine company to lead an additional 2½-inch line up the aerial ladder. This line never had to be charged; the first line had knocked down the fire, which hadn't extended much past the kitchen because of the building's first-class construction. Four firefighters had an extremely close call for what ended up being a one line fire.

My major error in judgment was entering the building in the first place. As Incident Commander, my duty was to direct the operations to attempt to locate and control the fire, evacuate the occupants, and try to protect the firefighters as they performed their duties. When the leader is inside the building and not in a command position, leadership is missing.

I suppose I could claim that because I was a new chief I didn't know any better, but I certainly had been to enough fires as a company officer to know the correct procedure for Incident Commanders. Maybe I wanted to express my eagerness, but I really expressed my stupidity. I feel fortunate that no one was injured or killed as a result of my extremely poor judgment in taking an elevator to the fire floor and not being properly equipped.

Learning from Experience

On the next tour of duty, I paid a visit to the Deputy Chief and suggested that we consider establishing procedures, including the use of elevators, for these taller buildings. The Deputy reported my concerns to Headquarters. He found out that, because so many of these buildings were under construction, the department was already starting to develop procedures for high-rises. My report did speed up these efforts, and soon afterward we started publishing information pertaining to these taller structures. We also started to drill fire companies in the districts involved, as well as those in outlying areas of the city.

Setting the Standard

During the next ten years, as many high-rise buildings were erected, the Standard Operating Procedures were expanded. In time, laws were passed in regard to control of elevators, and in the mid-1970s all newly constructed

"The Commish." Photo by Bill Noonan.

buildings exceeding 70 feet in height were required to have automatic sprinklers. The new sprinkler requirements didn't apply to existing structures, but after a major fire in the unsprinkelered Prudential Tower in 1986, a retroactive sprinkler law was enacted with a ten year deadline. By the start of the twenty-first century, this task was completed.

> *We must not be lulled into thinking we're safe.*

In spite of these fire protection requirements, we must not be lulled into thinking we're safe. The twin towers in New York that were attacked on September 11, 2001, were fully protected against fire, but both of them collapsed. Firefighting is a business of surprises.

> *Firefighting is a business of surprises.*

When firefighters are in charge of any incident, their primary responsibility is to protect the safety of the citizens and the firefighters. The Incident Commander should be in a clearly visible and accessible location, and should give orders based on what is visible, supplemented by information received from aides and company officers operating in interior locations that the superior is unable to observe.

I was a chief officer for twenty-six years and while there's no doubt I made many other errors during my career, I never forgot this frightening lesson, during my first year in that rank. I can guarantee you that I never made this particular mistake again. Firefighters who are determined to advance in the fire service would do well to study the experiences of those who have preceded them. Still, nothing compares to personal experience. Each time firefighters

> *Nothing compares to personal experience.*

respond to a serious incident, they should absorb every bit of information possible and write them down. Memories fade over time, but the written word remains.

Lessons Learned

- 🔥 **We should learn from our mistakes.** Even a legendary Fire Chief like Leo Stapleton admits to making mistakes. He took a negative and turned it into a positive. Beyond learning from his mistakes, he taught others to avoid them. After this high-rise incident, Leo Stapleton became one of the most knowledgeable firefighters in the United States in high-rise firefighting.

- 🔥 **We need high-rise training, training, and more training!** Every department that responds to high-rise buildings must train in this field. The study of risk management teaches us that the most dangerous incidents are *high risk and low volume.* Firefighters don't respond to a lot of high-rise fires, so they are low volume, but when we do they are always high-risk.

- 🔥 **We have to fully understand our department's SOP or SOG.** Firefighters must understand their department's Standard Operating Procedures or Guidelines. Just as all the members of successful football teams are required to understand the responsibilities of their teammates on every play, firefighters must see the whole picture and comprehend what everyone is doing.

Discussion Questions

1. What are the standard operating guidelines for high-rise fire-fighting for the fire department in your area?

2. What are the standard operating guidelines for this fire department concerning the use of elevators at fires?

3. What modern-day procedures and equipment can you name that might have saved lost firefighters in the past?

4. What other type of incidents does your department respond to that would fit in the dangerous "high risk/low volume" category?

Saving Lives

M y story is not about a particular incident; rather, it is about the last twelve years at Engine 16. It's not that there is a shortage of fire stories to talk about. It's just that I have a more important story to tell. Like most of the stories in this book, my story is also about saving lives—just in a different way.

Firefighter Kirkland
Flowers, Engine 16
15 Years in the Fire Service

First Day at Engine 16

From my first day at Engine 16, I felt at home. As I pulled into the parking lot, I looked around and smiled. I had just transferred from a small firehouse in a quiet area of the city. Engine 16's quarters was also a small firehouse, but that's where the comparison ended. This station was surrounded by several high-rises, all of which were low-income housing, called The Projects. Furthermore, this neighborhood had one of the highest crime rates in the city. People could set their watches by the gunshots. "Wow," I thought, "I finally had made it." I was on the busiest engine in the city, just where I wanted to be. Twelve years have passed since this day but I still get excited when I come to work. In the beginning, the firehouse doors were always closed, and we had to stay away from windows to avoid getting shot. Firefighters' cars were continually being broken into or vandalized. Gangs ruled The Projects, and shootings and stabbings were regular

> *I was on the busiest engine in the city, just where I wanted to be.*

occurrences. On any given school day, close to a hundred young kids would be running the streets and just hanging out. If you asked them why they weren't in school, they would just stare at you.

Opening the Doors and Our Hearts

I found out that an officer on one of the other shifts was working with some of the kids. I thought that this was a great idea and I wanted to join in. One day I asked my officer if I could open the firehouse door and invite the kids to come in. He agreed, as long as I would stand watch. As soon as the doors opened, the curious kids started to come in.

> *This marked the beginning of a relationship that grew with time.*

This marked the beginning of a relationship that grew with time, and we gradually gained their confidence. They wanted to hang out in the firehouse a lot, but we let them stop in only before and after school. Some of the kids were hungry, and we fed them. We started bringing in clothes for those who needed them.

Becoming Mentors

One day when some kids were hanging around after school, I asked them about their homework. Most of them turned to leave but others stayed. Some of us started working with any kid who would let us help. Eventually, when we went out on a call, we trusted the kids to watch the house for us.

When the gangs saw what we were doing with the kids, the break-ins and vandalizing of firefighters' cars started slowing down. Fewer young kids were cutting school because they knew we wouldn't let them visit us if they did.

We started giving fire department T-shirts and big pencils to kids who had perfect attendance at school. A ten-year-old named Griff still wouldn't go to school. We saw him hanging out almost every day. Like most of these kids, he had a rough home life, and he just didn't care about school.

We made him a deal. He had told me that he really wanted a bike, so I promised him that if he went to school every day for just one marking period, we would give him a bike. At this point, grades didn't matter, just attendance. In our agreement, he couldn't be tardy or have any absences for the entire marking period.

Within a short time the principal of the elementary school paid us a visit. The attendance at her school had been about 40 percent, and she started to notice a significant increase. She also noticed that some of the kids were wearing fire department T-shirts. What really shocked her, though, was that Griff was attending school every day and he was on time, after nearly a hundred tardies and absences that school year. She thanked us and encouraged us to keep on with our efforts.

The Payoff

The next marking period, Griff walked into our firehouse, report card in hand, grinning from ear to ear. The grades weren't the best—a couple of F's and a couple of D's, but also a couple of C's. Most important was his perfect attendance.

Griff kept up his part of the deal, and so did we. When we rolled out his "new" bike, he looked shocked. Actually, we had taken an old bike, bought new parts, and spent several hours making it look new. He was as excited as any kid I had ever seen.

When the other kids saw Griff's bike, they wanted bikes, too. We started FITCH (Firefighters In The Community Helping), making T-shirts with the FITCH logo and taking donations of bikes we could fix up. The first year we gave away forty bikes and hundreds of T-shirts and pencils. The school principal came to see us again, with the news: In a year and a half, the attendance at the elementary school had gone from 40 percent to around 94 percent!

Word spread quickly, and the principals of other neighboring schools came by to ask for advice with their attendance. We now work with fifteen elementary schools. FITCH keeps growing, and this year we will give away 5,000 bikes and 500 computers. Most of the bikes are donated and we fix them up. The Marriott Hotel chain gives us their old computers, and a group of sailors from the Great Lakes Navy Base helps us reprogram them.

We set up a computer lab at the community center next door with more than fifty computers. Our after-school tutoring grew so much that we had to move the program to the community center. Teachers and firefighters volunteer their time to staff the center.

Over the years we have had several fieldtrips, taking the kids on bus trips, roller-skating, to the beach, or downtown. We pass out consent forms for each trip, and some of them come back with notes saying, "I don't care what you do" or "Take him and keep him." For some of these kids, this is their first time out of The Projects. With gang problems and guns all around, they don't wander far.

We have rules that the kids must follow to visit the firehouse. They have to wear clean clothes, comb their hair, and brush their teeth. They have to go to school and do their homework. They have to go home at 7 p.m. on school nights and 9 p.m. on weekends. Most of them actually seem to appreciate having rules because ours are the only rules that some of these kids have in their lives other than at school.

We have gained respect from the gangs. When we respond to a call in The Projects, the people let us through and run ahead to lead the way. When a fellow firefighter's radio was stolen on my off day, a firefighter called me at home. When I walked into The Projects, I didn't even have to ask for it. One of the gang members handed it over and apologized for the guy who had stolen it.

We have gained respect from the gangs.

That isn't all. We deliver a lot of babies and treat a lot of injuries. One day a gang member walked into the firehouse with a knife sticking out of his chest and asking, "Where's Kirk?"

Another kid walked in and said, "I need Kirk. My grandfather is having a heart attack and he'll only go to the hospital with Kirk." I had to talk him into going with the ambulance. Over a period of time they have learned to trust all of our firefighters. Recently, some teenagers caught a guy stealing hubcaps from our parking lot and they held him until we returned from a call.

We deliver a lot of babies and treat a lot of injuries.

We Don't Always Win

Even though we have seen some positive results over the years, we also have seen some major disappointments. One boy we called Big Al had been hanging around since he was six years old. Both his mom and dad were on drugs. We fed him every day and worked real hard with him, but we couldn't keep him in school. He was selling drugs and was killed before he was seventeen. Another one of our boys couldn't stay away from drugs or the gangs. As he was talking to one of our firefighters in front of the firehouse, gang members shot him—dead at age sixteen.

Our Reward

We do have our share of happy stories. Remember Griff? He was the first kid we gave a bike to, and one of our greatest accomplishments. He stayed off drugs and out of the gangs, and now he attends Northern Illinois University and is doing well.

Another one of our kids, Samantha, was so smart that in sixth grade she was writing children's books about life in The Projects. We kept her supplied with pens, paper, and notebooks, and she wrote the books and sold them for 10 cents each. We told her that if she kept it up, she would be a famous author some day. Today she is in college, with no babies, no drugs, and a publisher seriously looking at her work.

At least four more of these kids are in college right now, and several more will start college this year. Each of our firefighters who gave their time, sweat, and heart to these kids is reaping the reward. For some of these kids, this is their only chance in life. My only regret is that I wish I could do more.

For some of these kids, this is their only chance in life.

Lessons Learned

- **There is more than one way to save lives.** Kirkland Flowers, and the crew of Engine 16 remind us that firefighters are role models. We have a chance to help shape and mold the futures of children throughout the country. Kids look up to us, and we must never let them down.

- **Reach out to your community.** Stay in tune to what is going on in your area. Every community has its problems. Is there something that we as firefighters can do to help? The work may be hard but the rewards are great.

Discussion Questions

1. What could you and your fellow firefighters do to make a difference in your community?

2. How does a community gain from a firefighter mentoring program?

3. What benefits does the fire department receive for such programs?

Saving Lives: The Instructor

Battalion Chief
Steve Chikerotis
27 Years in the Fire Service

Fighting fires can be quite challenging. The accomplishments we achieve along the way are what fuel us. There is another part of my career that I always found rewarding—"I am a teacher." Ever since the early years of my career when I became a student of the game, I have shared the lessons that I learned with other firefighters. This is called "giving back to the job," and this is the responsibility of all firefighters. Sharing our knowledge with others will save lives over the years. Teaching is a big part of both a company officer's and a chief officer's duties. To be a good officer you must be a good teacher; it comes with the territory.

Throughout my career I also have spent time as an instructor at the fire academy, teaching in-service classes, and also training new recruit classes. Taking raw recruits and developing them into firefighters is a challenging task, but I believe it is very important that the recruits have a strong foundation on which to build their careers. Over the years I have been involved with hundreds of new recruits. This story is about one of them.

In May of 1990, I left my assignment as a lieutenant in the field to take a three-month detail as an instructor of 100 recruits—whom we call "candidate firefighters." The instruction would involve classroom, hands-on, and physical fitness training.

Time to Share Knowledge

The men and women of this class came from a variety of races and nationalities, and there was also a wide range in educational background.

Despite the differences, these candidates all seemed to have a good attitude and a willingness to work as a team. I looked forward to the next three months.

One of the candidates really stuck out and quickly became one of my favorites. He was an African American in his late twenties with an outgoing personality. He was in superb physical shape and could crack walnuts with his handshake! The enthusiasm he brought to the class each day was a breath of fresh air. Early on, though, he told me that he was having a hard time studying because he had been out of school for so long. I told him not to worry about it, just give 100 percent, and I would help him.

As time progressed, all of the other candidates seemed to be way ahead of him academically. He was one of the best in both the hands-on and physical fitness training but those skills didn't prevent him from failing miserably on the tests. At our weekly meetings about the students' progress, our chief expressed concern about his seeming inability to learn, and that his scores wouldn't enable him to pass the required State Firefighter II exam at the end of the three-month class. He was considering letting him go.

I argued strongly on his behalf, telling the chief that, besides the candidate's physical prowess, I saw something special in him. He had a great attitude and a huge heart—traits we can't teach. Further, I had a feeling that he wasn't going to be a good firefighter—he was going to be a great one! I requested a little time for my candidate to show that he could raise his academic scores. The chief agreed to a brief extension but advised me that we were on a "short leash." Immediate progress needed to be demonstrated.

He had a great attitude and a huge heart—traits we can't teach.

After the meeting, I went straight to my candidate and told him what we were up against. I also told him that I believed in him and as of that minute, I was adopting him! I would be his new step-dad. We would meet two hours early every morning and stay two hours late each night to study.

His reaction confirmed my belief in him. He didn't look scared or upset, he looked determined. He was a fighter and wouldn't be derailed in achieving his goal. He grinned, shook my hand with that walnut-crushing grip and said, "When do we start, Dad?"

Over the next two-and-a-half months, we met every morning and every night as agreed. His progress was slow but steady, and his enthusiasm was my reward. As the date of the state exam approached, other students joined our study group. My new "stepson" started meeting with other candidates at lunchtime, and I burst with pride as they started to turn to him for help. He was a natural leader, and his peers picked up on that trait.

When the day of the Firefighter II exam rolled around, we didn't fear it. We looked forward to it. Watching through the window during the 400-question exam, I felt like a proud father as my candidate sat there confidently answering the questions.

When the results came back from the state, not only had he passed but so had the entire class. The whole class had caught his contagious enthusiasm and it had worked.

Fire service instructors and officers should always look for the diamond in the lump of coal. Great rewards for yourself and your department can come out of this. How has my candidate done in his career? You tell me. His name is "Firefighter Kirkland Flowers," and you have just read about him in the previous story. I cannot imagine our fire department or our great city without this man. To this day, Kirk calls me "Dad" and greets me with a hug, and I could not be prouder of his accomplishments.

Lessons Learned

- **Give back to the job.** The goal of all firefighters should be to give back to the job and leave the fire department and fire service a better place than they found it.

- **Education saves lives.** We should strive to be students of the game and teachers throughout our career. A football team could never win a game without practice, and a fire company shouldn't expect to be successful without training and drilling.

- **A good attitude is contagious.** A winning atmosphere at work starts with you. If you treat others like you would like to be treated, you will see the results. Take pride in yourself and your work.

Discussion Questions

1. What are some ways we can improve our learning?

2. What benefits besides education does a group get from training together?

3. How can you as an individual make the lives around you better?

A Tough Decision

Retired Chicago Fire Commissioner Raymond E. Orozco, Sr.
38 Years in the Fire Service

Assistant Deputy Commissioner Raymond Orozco, Jr.
26 Years in the Fire Service

The following are two accounts of the same event by father and son firefighters. The stories are written in the first person, based on interviews with former Chicago Fire Commissioner Raymond E. Orozco, Sr., now retired, and his son, Assistant Deputy Fire Commissioner Raymond E. Orozco, Jr. The two accounts are differentiated by their ranks at the time—Ray, Sr., as Commissioner and Ray, Jr., as Lieutenant.

Lieutenant Orozco

On that cold December day, the wind seemed to cut right through a person. We were sitting around the firehouse that evening waiting for dinner when I noticed that the activity on the fire radio was really starting to pick up around the city. As lieutenant of Engine 23 that day, I announced, "Saddle up, boys, I think it's about to start."

Sure enough, a second later the speaker crackled, sending us to a fire in a high-rise, low-income housing project. As we ran to the rig, the firefighters yelled, "What are you, psychic?" "Forget predicting fires, how about tonight's lottery numbers."

We were out the door in mere seconds. Steam was rising out of the sewers as we made our way through the traffic.

Within a few minutes we were lugging hose and equipment up to the fire apartment, which was on the seventh floor of the high-rise. The fire wasn't much to speak of, just a routine one-line fire. While we were overhauling (extinguishing the last of the fire such as smoldering materials, and searching for hidden fire behind walls and ceilings), I heard a truck company report on the scene of a church fire a couple of miles away from us.

I stood on the balcony and looked over in that direction. In the distance a thick column of black smoke rose in the night sky.

"We might end up on that one," I muttered, pointing out the smoke plume to those around me.

By the time we got back to our pumper, I heard the chief pull a second alarm, two more engines, another truck, and another chief. I grabbed my radio and told the alarm office that we were available and asked if they wanted us to take it. The voice at the other end told me they had it covered and didn't need us on this fire.

For some reason, I had a gut feeling that this fire was going to escalate. I told the engineer (driver and pump operator) to start driving in that direction. "We'll take the scenic route back home just in case they call us," I said.

We were about four blocks away from the fire when a voice screamed over the radio: "Emergency! Emergency! There's been a collapse! We've got firefighters down!"

We threw on our lights and siren and started in, as the chief requested a 2-11 (a third alarm) and an EMS Plan I (an Emergency Medical Service call for five ambulances). By the time the alarm office was calling for us to respond, we were already pulling up to the scene. All that remained of a huge church building was the rear wall, which stood about 60 feet high. The wall, which had supported the gable roof, now appeared to be pointing to the sky and was leaning perilously inward. All that remained of the sidewalls was an 8-foot wall that circled the perimeter. Jutting up over the short sidewall was what appeared to be the butt end of the collapsed roof rafters. I would find out later that it actually consisted of the butt ends of the floor joists, which had collapsed in the center and formed a V-shape.

"I'm missing my company!"

Pockets of fire were everywhere. Thick smoke was hanging close to the ground as it poured out of every opening. Then the officer from Engine 107 staggered out of the smoke, bloody and dazed and missing his helmet. His voice shaking, he said, "I'm missing my company!"

One of the chiefs started to organize search groups. With my company and Truck 7, we climbed a short ladder over the 8-foot-high remains of a wall and slid down into the rubble. Soon the lights of several firefighters dotted the smoky rubble as they joined the search.

What Happened Before We Arrived?

Before we got there, Engine 107 had led a 2½″ line through a door on the front corner of the church. The firefighters worked their way deep into the church when it collapsed suddenly with little or no warning. Just prior to the collapse, they saw a lot of fire high above them, and also on a back wall to the left of the altar, but they were still able to stand up because the heat wasn't that bad. The smoke was banking down and starting to obscure their visibility just before the collapse.

As the building started to collapse behind them, those who survived ran toward the altar and escaped out a back door. Another company, Engine 25, also was inside the church with a 2½″ line. When the building collapsed around them, these firefighters were able to dive out the front door. They had been under the choir loft, which shielded them momentarily so they could escape.

What Happened After We Got There?

Now we found ourselves slipping and sliding in the debris looking for an unknown number of firefighters. Fires burned below us, and smoke obscured our vision. The floor angled in toward the center like a big V, making it tough to stay on our feet. Smoke was pumping out of holes all over the place.

> *The dangerous back wall loomed high above us like a giant fly swatter.*

The dangerous back wall loomed high above us like a giant fly swatter. One of the firefighters got some lifelines, and I hooked one to my belt. We tried to squeeze into a hole, but the heavy smoke resulted in zero visibility. We could hear the debris crashing around us. Then the message came over the radio: "Back out! Everyone back out for a roll call."

Regaining Control

The roll call gave the chiefs a chance to regain control of the incident. After an accountability check with each company officer we now knew what we were up against. We were missing one firefighter, and we knew where he was last seen. The missing firefighter Calvin was assigned to Engine 107, and he had been on the nozzle of the attack line at the time of the collapse.

One of the chiefs asked a tower ladder to raise up its basket and pull down a dangerous section of the rear wall. With the rear wall somewhat stabilized, we returned to our search, with three search teams of four firefighters each. The first team went to the rear and tried to gain entry to the basement. A second group attempted to gain entry through the front door. My group tried to follow Engine 107's hoseline through the front corner door. About 10 feet inside the door the hoseline was buried under a mountain of debris. We started to dig by hand through the rubble.

Commissioner Orozco

I was notified immediately after the collapse. My driver picked me up at my home, and we responded. When we arrived, I did a quick survey and studied the situation. The church had totally collapsed, and several large sections of wall were leaning precariously. Fire was still actively burning in the basement and throughout the collapse area. A two-story school next to the church also was on fire.

Several firefighters were actively searching for our missing firefighter. My main concern now was *time*. By now, Calvin had been missing close to 40 minutes. Under these conditions there was no chance that he was still alive, and my total concern shifted to the safety of my remaining firefighters. It was clear that the rescue effort had been converted to a recovery operation.

Worried about a secondary collapse, I decided that the rescuers were in harm's way. I ordered everyone out until we could knock down some of the fire and reevaluate the situation.

Lieutenant Orozco

The order came over the radio for all of us to evacuate the building.

"No!" I couldn't believe it. "We can't back out with our brother in here." Thinking we had no choice, we backed out, against my wishes. As soon as I came out of the building, I saw my dad and realized that he was the Incident Commander. I felt sorry for him, knowing what a tough decision he had just made. I asked my whole company to walk to a pay phone down the block and call their family to let them know they were safe.

My crew and I stood and watched the fire streams from hoselines and master-streams devices (large water streams) directed at the huge pockets of fire. From this vantage point, I began to realize that we had lost a brother firefighter tonight. After several minutes, most of the fire was knocked down. After the streams were shut down, we were allowed to make another attempt at recovering our brother. Each team had a specific area to search.

Commissioner Orozco

During one of the recovery operations, I entered the collapse area to check firsthand on the conditions. A group of four firefighters had formed a chain to pass debris away from a search area. One of the firefighters turned to hand over some bricks, and I took them from him. When we came face to face, I realized that it was my son. I knew he was working on the scene, but it was the first time I had seen him, and I could see the pain in his eyes. I told him to keep his crew safe and went off to check on the other teams.

A short while later the fires started flaring up underneath us again. Once again I had to pull everyone out of the building.

Lieutenant Orozco

We were backed out of the building for a third time, and again master streams were put to work. For the first time, I realized how cold I was. I sent my company to a warm-up bus to thaw out, and I saw an opportunity to talk to my dad alone for the first time. When we were working, he was our Fire Commissioner, but when we were alone, he was my dad. I wanted to make sure he was okay. "How's it going, Dad?"

"Not good, Ray. We lost a good man tonight." I winced at the pain in his voice, and it looked like he had aged five years that night.

Commissioner Orozco

Throughout the long, cold night we fought the fire and made many attempts at recovering the lost fireman. I had to rotate fresh firefighters in, but I didn't strike out the 3-11 (a fourth-alarm fire) until we finally recovered our firefighter's body just after sunrise.

Lieutenant and Commissioner Orozco at fire.

I had been the Fire Commissioner for only about six months, and this was the first firefighter we had lost in that time. Prior to my becoming Commissioner, we had a few bad years when we lost five or six firefighters lives per year. I remember standing there thinking, "Please, God, don't let this happen again."

Lieutenant Orozco

We were sent home from the fire around three in the morning. None of us wanted to leave, but we were no good to anyone at that time. We had been there around nine hours. We were frozen solid and so was our equipment. Back at company quarters we were able to thaw out. A few hours later the next shift came in and relieved us, but none of us went home. Our whole group grabbed our frozen gear and headed back to the scene in our own cars to help out. Shortly after we arrived, his body was found buried under debris in the twelve-foot deep basement. Although I didn't know Calvin personally, he was our brother, and we did everything we could have done to save him that night. God bless this hero.

Lessons Learned

- **High ceilings offer a false sense of security.** High ceilings allow the heat and smoke to rise high above the heads of firefighters. Most firefighters associate danger with high heat and zero visibility and are lulled into a false sense of security when these conditions are not present. We must make sure that we see the whole picture.

- **We have to read the smoke conditions.** A high volume of smoke means a high volume of fire. Darker, fast-moving smoke is hot. Dark brown smoke is wood burning or structural involvement.

- **We must think defensive.** While fighting church fires in most cases, if there is too much fire for two 2½" lines to quickly extinguish, we should think defensive. The first line is directed at the seat of the fire and the second line is directed above. Churches are usually fast-spreading fires.

- **Good communications is essential.** Teams outside of a high-ceiling building often can tell conditions are bad. They must warn the interior companies, who often can't tell the severity of the fire due to poor visibility and low heat conditions at the floor level.

- **We must beware of secondary collapse!** Collapse weakens the structure and secondary collapses are inevitable. The impact load of firefighters working can trigger these collapses. Monitor what is above and below you at all times.

Discussion Questions

1. How do high ceilings make the fireground more dangerous?

2. In a similar situation, what indicators of collapse can we look for?

3. How could the proper use of communications improve firefighters' safety prior to collapse?

And I Thought Fighting Fires Was Hard Work...

Chief of Department,
Joanne Hayes-White
San Francisco Fire
Department
16 Years in the Fire Service

We sprang to our feet as Chief Postel strode up to the podium. How regal and powerful he looked in his black, double-breasted dress coat, the gold badge and braids dancing against the rich, black fabric. Here was the Chief of the San Francisco Fire Department, welcoming the Fire Academy class on our first day. We stood proudly, clad in our navy blue jumpsuits and holding tightly to our yellow probie (new recruit) helmets. You could have heard a pin drop as the Chief spoke of pride, honor, and dedication. At that moment I knew I had chosen the right profession. The most vivid memory of all was how every member of the training staff, men who terrified and intimidated everyone in our class, seemed in awe of "The Chief."

From Recruit #10 to Chief of Department

Fast forward to January 2004. Nearly fifteen years had passed since my humble beginning that first day, when I was known simply as Recruit #10. Now feeling like I was in the middle of an unbelievable dream, Gavin Newsom, the Mayor of San Francisco, was swearing me in as the new Chief of Department. Although I had the same feelings of pride and excitement that I felt that first day, I had acquired an added dose of humility, and more than a healthy dose of anxiety and fear.

When I look in a mirror I see the same person. I'm still Joanne, a woman who grew up in this city, and has always felt an emotional connection to our citizens. I also grew up having a deep respect for the San Francisco Fire Department and the men and women serving in its uniform. These attitudes may well be my best asset for the job. On the day I was sworn in as Chief of Department, I made a personal vow to re-instill that same pride and respect in our members, and in the public we serve.

On the day I was sworn in, I knew I was being thrown into deep water. Even though our Department has a long and rich tradition of service to our city, I was being asked to assume command during one of the most difficult periods in the department's history. The integration of Emergency Medical Services (EMS) into our Department had been painful and was in need of leadership and vision.

September 11, 2001, had suddenly forced the Department to integrate a new level of education and skills into its repertoire. The city was facing a high-dollar budget shortfall that promised to get worse. On top of all that, our Department had been the target of numerous reports and audits that severely criticized nearly every aspect of our operation. Worst of all, these problems had taken their toll on the members of our Department. Although we all remained dedicated and committed to serving the citizenry, morale was at an all-time low.

My work certainly was cut out for me. Was I prepared for the task? Had my training and experiences as I rose through the ranks prepared me for these challenges? Looking back, I proved to be up to the challenges I faced in this position, and I owe this to the on-the-job education I received every day of my career. Every experience, from our first day of training to our last day on the job, shapes and molds our careers. Furthermore, those who actively pursue education can speed up the learning curve.

The Journey Begins

The journey began on that day when I walked into the Fire Academy. Like any young recruit, my pride and excitement mingled with a healthy dose of fear. The training at the Fire Academy gave me a basic understanding of the job. This knowledge and the skills learned during the training prepared me to fight fires, render aid to the sick and injured, and embark on search and rescue missions. The most important thing I learned in the Fire Academy, however, was to trust and cherish the men and women I work with.

My pursuit of education took me to the busiest fire stations in our city. From experienced and knowledgeable firefighters, I learned how to apply the skills I had learned at the academy, as well as the lessons a person can't get from any textbook. I also learned about compassion. When people call the Fire Department, they need our help, and they deserve our best efforts, our kindness, and, above all, our respect. I learned to treat all people as if they were family members.

On-the-Job Training

I remember my first greater alarm fire (large fire) like it was yesterday. I was still a new recruit, a probie. The firefighters in Station 1 were a tight-knit group. They had been together for years and had been tested again and again on the "field of battle." They stood shoulder to shoulder, fighting fires and losing brothers in the line of duty.

Breaking into that environment wasn't easy. I had to suffer through the initiation rites that every new firefighter must endure when assigned to a busy fire station. It meant doing a disproportionate share of dishes, housework, and cooking. The few times that someone would show sympathy for the probie, it would soon be lost in the "tough as nails" group dynamic of the "Alley Cats." I felt like an outsider, a stranger in a strange land. That was all about to change.

On the night we pulled up to a five-story, low-income hotel fully involved with fire, all the ribbing and hazing fell away. In that moment, the real magic of the Fire Service and the men and women who serve became a reality. The firefighters took me under their wing. Knowing that I was a novice, they watched out for me like a member of their family. Make no mistake—I was expected to pull my weight. When I didn't balk or hesitate at going in, I became a part of something much greater than its parts. I was a member of a team, a firefighter! On that night I felt like I had become a member of a family.

As I moved through the ranks, I took all those lessons and values with me. I found myself in a position to mentor other young firefighters, sharing as much as I could with them. This vocation also gave me a new appreciation for family and friends, having seen firsthand how fragile life is, how common it is for someone to suffer a heart attack, be involved in an accident, or fall to a stray bullet. Witnessing so many tragedies has taught me the most important lesson of all: Never take the time you have with your family for granted. Kiss and hug your kids every chance you get—even if they act like they hate it!

From the Field to an Administrative Position

After nine years in the field, I found myself in my first administrative position, supervising the Communications Center. The men and women I worked with there were no less dedicated, no less committed than our brothers and sisters in the field. The difference was that we were no longer calling upon our skills and courage at emergency scenes. We were the faceless voices that elicited information, provided instruction until help arrived, and tried to reassure callers that things would be okay. We all felt firsthand the frustration of not being able to be there, but we also knew how important our jobs were.

In supervising the Communication Center, I had to master the nuances of scheduling, recognize the harsh reality of budgets, and wrestle with all the responsibilities that come with upper management. At first I felt awash in a sea of details, frustrated that nothing in the Fire Academy or in the field had prepared me for this new position. Soon I realized that the skills and lessons I brought with me from the firehouse really did translate to this administrative position. I recognized that as firefighters, we learn how to think on our feet, how to prioritize, how to immediately focus on what is critical to our mission, and to save the rest for later. I knew that forming a team, shaping a group of individuals into a cohesive unit, is as critical for communications as it was on the fireground. Above all, I never forgot that the well-being of the people under my command has to come first.

On the way up the ladder, I served in a few other administrative positions. One of these was Assistant Deputy Chief over the Division of Support Services, which was responsible for Department infrastructure. This division included the auxiliary water system, supply system, apparatus, facilities, uniforms, and Personal Protection Equipment, which exposed me to a completely different side of the Department.

Among my other assignments, I was charged with oversight of recruit and inservice training, and I served as the Director of Training. This role was a tremendous challenge but also a good fit for me, as I firmly believe that training is the foundation for the success of any organization. In the arena of Fire Service, this could not be more true. I prided myself on developing innovative and cutting-edge curriculum and training modules. It is a prerequisite for officers and chiefs to be good teachers. The lessons we learn on this job just might save our lives.

The lessons we learn on this job just might save our lives.

Linking Past and Future

Our Mayor has been unwavering in his support, and for that I am grateful. I was given free rein to choose the members of my command staff, and collectively we are shaping a new vision and direction for our Department. The course we have plotted is challenging, innovative, and ambitious for a Department so steeped in tradition. While we're committed to moving the Department in a new direction, we're just as committed to holding on to the history and traditions that make the San Francisco Fire Department one of the best.

We're reconfiguring the way we deliver EMS services. We've introduced modern management techniques, including data analysis, performance management, and customer service performance benchmarks. We've made accountability a tenet in our administration, and reinstituted public recognition for exemplary and heroic performance. First and foremost, our commitment is to each other. We remind ourselves that the strength of our Department,

We are a family.

its heart and soul, is the oath that we all took when we became firefighters. It joins us in a way that can't be quantified or underestimated. We are a family.

Not all of my duties as Chief have been pleasant. I have had to move forward with disciplinary actions, some of which resulted in the termination of members I consider friends. Because of budgetary demands, I've become more familiar with saying "no" than I ever anticipated, often for projects or ideas for which I otherwise would have been an advocate. Most difficult of all, I've found myself at the hospital bedside of Department members who have been gravely injured, both on and off duty.

Once again, I come back to the value of the training I received in the Academy and honed on the fireground. From that base, I learned how to prioritize the critical from the not so critical, when to stand and fight, when to fall back and regroup, and how to form a team that knows it can rely on each and every member.

My journey hasn't been easy, but it has been rewarding. We remain steadfast in our resolve never to compromise on our ability to carry out the mission of the San Francisco Fire Department—to protect the health, safety, and property of the residents and visitors of our great city—and we will.

Lessons Learned

- **Advancement requires preparation.** Everything we do shapes and molds our future. We must always be students of the game.

- **We must set goals,** write them down, and then pursue them.

- **We have to be aggressive in the quest for education** and never shy away from challenges.

- **We must never forget where we came from.** In the Fire Service, the most important resource we have is our people. The welfare of the people comes first. When we take care of our own people, it usually pays dividends.

- **We must remain inquisitive.** This means that after you've learned your job, you should take the time to learn everything you can about it. Always ask, "How could I do this better?"

- **Teamwork and trust are essential.** Teams do not form by themselves. Leaders must build them.

Discussion Questions

1. How are you preparing yourself for growing in your chosen career?

2. Can writing down your personal goals help you achieve them? If so, how?

3. What are some ways you could put yourself in the "right place at the right time" for advancement?

Prepare Your Troops for Battle

L adder 176 is known as the "Tin House Truck," for good reason. The first fire station was a temporary building constructed of pre-fabricated metal—hence the nickname. This unit was busy enough and experienced enough to be considered a choice unit to work for. Having worked alongside this unit for many years, I was familiar with the officers and personnel who made up the unit and was fortunate to be assigned there as the captain.

Retired Captain
John Vigiano,
Fire Department of
New York
39 Years in the Fire Service

Learning Humility

The 1990s brought a new innovation into the staffing of units, called a rotation system. The probies (new members) would be assigned to a unit for one year and then would be rotated out to another unit for the second year, and again for the third year. At the conclusion of this cycle, they would come back to the unit they started with. On paper this sounded simple enough. Besides, it gave young members an opportunity to work in other areas of the city and experience methods of operation there.

As the captain, I faced somewhat of a challenge in the cockiness of the younger members of the unit. Those five-year firefighters thought they knew everything but hadn't really seen the fire duty of the more senior members—especially those of us who worked during the late 1960s and early 1970s. I had to find a way to take the wind out of their sails without jeopardizing the spirit that makes them good firefighters. Cockiness is part of it, but it has to be tempered.

Fortunately, along with the rotation program, an administrator in headquarters came up with the idea of classifying units as "A," "B," or "C" companies. "A" units were the busiest, "B" not as busy, and "C" the slower units. We derived these classifications from numbers the units submitted every year based on number of runs and "workers," or working fires.

A few months before this program became official, I received a heads-up on the classification and changed Ladder 176 from an "A" to a "B" category. When the list came out officially classifying the units, the young firefighters went nuts.

Comments started flying around the kitchen: "Who is Headquarters kidding? . . . We're a busy unit. . . ." "How can this be? We're in the 15th Division . . ." (all the ladder companies in the 15th Division had been designated "A" companies—except Ladder 176). Then the phone calls started coming in from the surrounding units, offering their "condolences." Their words of sympathy drove the knife in deeper.

This caused quite a stir in the house. The junior firefighters went to the senior firefighters, who in turn came to me for answers. I called a company meeting and offered, "Maybe we're not as busy as we thought. We should start the New Year a bit more humble. Numbers don't make a good unit. Firefighters make a good unit."

Ladder 176 went through the first year of the rotation as a "B" company. The lesson? Humility makes the tough even stronger.

> *Humility makes the tough even stronger.*

Balancing the Manpower

A second challenge that I had to deal with involved the new people who were being assigned to our unit. Because Ladder 176 had such an excellent reputation, many of the bosses' sons were being assigned to us. The word was out: To get into Ladder 176, they had to have a "hook" (clout). This wasn't a problem because they were all good guys, but our roster was overloaded and the firefighters were being detailed out on every tour to balance the manpower in the units that didn't have full rosters.

To compound the situation, the probies couldn't be detailed, so my senior firefighters were leaving every tour on details. This left the younger, less experienced members to take care of the unit. In Ladder 176, the junior firefighters invariably stepped up and took the detail for the more senior members, but that didn't solve the dilemma. It just helped a bit.

This dilemma was a double-edged sword. New members were being assigned to us every couple of months, and this gave us an opportunity to train more on the basics, which gave everyone an opportunity to refresh themselves. At the same time, the older, more experienced members had to be on top of their game, as the younger members were learning and asking questions every day.

Teaching Our Peers

The senior members started to run their own drills, and we instituted a training program. The seniors were to run a drill each day, and the younger crewmembers were to do the same thing, both groups teaching to their ability level. The younger members had to teach other young firefighters what they had learned during their first six months. This reinforced their level of knowledge and pleased the older members because their probie could teach other probies. A circle of pride began to develop.

With this new situation came another challenge. With a normal tour having a newer member (less than a year) and a probie (less than six months) working, the officer had the responsibility to protect them and yet maintain the efficiency of the unit. My policy was not to have two junior firefighters working inside with the officer. I wanted a more senior member working the "irons" (flat axe and haligan bar), as part of the inside team. This team consisted of an officer, a firefighter with the "can" (5-gallon hand pump for extinguishments), and a firefighter with the "irons." The primary duty of the "irons" firefighter is forcible entry.

This put the newest member in an outside position, usually the roof position. The newer members would be trained on the day tours by sending them to their position on every call. By going to the roof even on false alarms, the newest members would gain valuable experience that made them more comfortable at actual fires. From this training, along with the normal drill we held during each tour, they became fairly effective. My senior member usually was the chauffeur (driver), who was aware of the experience level of the member assigned to the roof position. In an actual fire, the chauffeur would advise the other ladder unit on the scene. The senior member and the second ladder company would keep an eye on them and evaluate the newer members' progress.

Going above the fire is an extremely dangerous assignment, so I instituted a policy to have my FE (forcible entry) locate our secondary means of egress as soon as we gained entrance to the floor above it. The firefighter with the "can" and the officer would conduct the search for life until the FE firefighter, using forcible-entry tools, could secure the way out. Then he would join in the search. Finding a secondary means of egress meant locating a window with access to a ladder and having a fire escape or a set-back if he were to get cut off from using the interior stairway.

If a fire escape were the only way out, the FE was to remove the entire window, including the sash (frame of double hung windows). This way there would be nothing to impede our exit. To ensure safety, the firefighter would have to leave the cross-sash in place of other windows that were vented to block the opening. By leaving it in place, in a high-heat, heavy-smoke condition, the firefighter would hit the sash. It would stop him just long enough to realize, "This isn't the way to exit!"

Good communication is essential to safety. After locating a secondary means of egress, the firefighter must communicate its location to all firefighters in the area. We are all responsible for knowing where we

> *Good communication is essential to safety.*

are at all times in relationship to means of egress. Our motto was, "Be smart and stay safe."

Lessons Learned

- **A fire company officer wears many hats.** An officer must be a blend of coach, counselor, teacher, tactician, disciplinarian, psychologist, and administrator. "In battle," the officer must be a leader.

- **Officers should pattern themselves after someone they admire.** John Vigiano provides a good example.

Discussion Questions

1. What are some ways you can prepare to be a leader?

2. Do you think an effective leader must also be a good role model? State your reasons.

3. What do you think are the most important traits of a strong leader?

Your Partner Can't See

Deputy District Chief
Jerome Shelton
25 Years in the Fire Service

My twenty-five-year career has been filled with experiences, some good and some bad. The important thing is that we learn from each and every incident. The incident in this story happened a couple of years ago, when I was the captain of Tower Ladder 34. Up to that point we were having an average day. We had responded to eight runs, but they hadn't required much work.

A Coach House Fire

At 1:00 a.m. we responded to a fire in a coach house, a smaller house at the rear of the lot. As we pulled up to Sector 1, which is the front of a building, the chief who had arrived ahead of us, called on the radio with urgency in his voice: "Tower 34, I need a primary search on the first floor!"

I grabbed our thermal imaging camera (TIC), and we quickly headed toward the rear of the building. This technology works by temperature differential and allows you to see in a smoke blackened room. Even in zero visibility conditions, when you look through the camera the hottest surface in the room appears white and the coolest surface black. As we walked up, I read the conditions—a 1½-story frame house, with heavy dark smoke belching out of the door to an enclosed back porch. The chief was helping a woman who had just jumped out of a high first-floor window. She was naked and scraped up from her 10-foot jump. A neighbor was handing the woman a blanket over the fence as we arrived.

"She said her husband is right inside that bedroom window," the chief yelled, pointing to a side window. "Their kids are already out," he added.

The engine company was just starting to stretch a line to the fire. I instructed a couple of my firefighters to assist them, and I took one with me to search. My partner, a guy called Dugey, was a tough, experienced firefighter. We worked well together. As we entered the enclosed porch, heavy smoke was pouring out of the basement door. I knew the engine was just behind us, so we hustled up the stairs into the heat. The door at the top of the stairs was still locked, and the heat was unbearable. Using his axe as a battering ram Dugey made quick work of the door, and we dove inside.

Once we hit the floor, the conditions were much more livable. I hit the sleep button on the TIC, and the LED screen came to life. Turning to Dugey, I asked, "Are you ready?"

He was, and we started to the right. The smoke was thick, but the TIC gave me good visibility. Without the camera, Dugey saw only darkness, so he kept his right hand to the wall. I held the TIC to my face, and the infrared technology allowed me to see the clear outline of a cluttered kitchen. The ceiling was white from heat buildup, and the floor, which was heated by the basement fire, was also white. With my newfound vision, I quickly navigated the cluttered kitchen and I headed down the hallway that led to the bedroom.

Separated from My Partner

I was already a couple feet into the bedroom when I realized that Dugey wasn't with me. I turned and gave a muffled yell through my mask: "Dugey, where are you?"

He didn't answer. I figured he must have been searching the other room, so I continued my own search. I squatted in the doorway and swept the room with the camera. Holding the LED screen close to my face, I could see the outline of the room, a small bedroom with a large, king-sized bed. The clutter left little space to move. As I made my way around the crowded bedroom, I kept asking myself, "Where is he?" Finally, about five feet from the window, I felt something between the mattress and the wall. It was our victim! He was invisible to the camera under a bedspread, which shielded his body heat. "Dugey, I got him!" I shouted.

No answer. This guy was huge, and he was squeezed into a tight space. I tried to move the bed, but it wouldn't budge.

The Rescue

Suddenly I heard breaking glass, followed by the sound of the window frame being cleared out and a shout of "Cap!" It was Dugey. What a welcome voice!

"I got him, Dugey. Stay on the ladder," I yelled, knowing that there was no room for both of us.

> *I had my work cut out for me.*

I pulled on the victim's arm, but he was wedged too tightly to move. I had my work cut out for me. I reached into my pocket and pulled out my rescue webbing, 16 feet of 1-inch webbing tied into a loop with a water knot (a non-slip knot used to join two ends of webbing together). I had to lie across the victim so I could force the loop around his waist, and then I clipped it to my belt with a carabiner. I stood up and leaned back, pulling with everything I had in me.

It felt like I was moving a building as I inched the guy toward the window. Once he was under the window, I reached under his armpits and lifted his upper torso into the sitting position. Then, from behind him I reached deep under his arms and lifted with my legs. As soon as his chest reached the windowsill, Dugey grabbed him. He and Charlie, another member of our company, carried him down the ladder, where an ambulance crew was waiting with a stretcher.

Exhausted, I was greeted by another crewmember who had come looking for me. We worked our way out of the building. By now the engine company was in control of the fire, and the conditions were improving. The cool night air was welcome as we tore off our masks and opened our coats.

After loading the victim into the ambulance, Dugey and Charlie met us by the truck. "You left me, Cap," Dugey chided.

I told him it was my fault and we would talk back at quarters.

By the time we got our equipment cleaned, it was close to 3:00 a.m. We all met in the kitchen for a quick talk. I apologized to Dugey for taking off on him. We have had the TIC only a short time, and I forgot that my partner didn't have the same vision device. Being able to see through smoke enables a person to move much faster than a partner.

Maintaining good communications between partners is always important and I neglected to do so. I was distracted due to the urgency of the situation and the use of this new technology. We would have to train more with the TIC so we could be proficient with it. Dugey said that when he lost me, he was having a hard time finding his way through the clutter so he backtracked to the door and helped Charlie with the ladder. He expected to meet up with me at the window.

This was good thinking on his part. Our company had just had RIT training at the Fire Academy. We drilled on removing victims from tight spaces, and how to use rescue webbing. This training really came in handy, and I assured my fellow firefighters that we were going to continue to train hard. On this night it was obvious that all of our hard work had paid off. We train to be the best, and I couldn't be more proud of my firefighters.

Lessons Learned

- **Training is the answer.** Captain Shelton was able to utilize Rapid Intervention Training (RIT) that he had just received to accomplish the demanding rescue. Training every day is essential to stay on top of our game.

- **We must all know how to use thermal imaging cameras (TIC).** Although TIC technology had been around for a few years, it was still new to our Department at the time of this incident. TIC is a valuable tool that can save lives, but like any piece of equipment, it has its strengths and limitations of which we have to be aware.

- **What you don't know can hurt you.** Tower Ladder 34 found out that we needed more TIC training, and began working on it every day until we felt comfortable using it. Taking technology for granted can jeopardize the safety of firefighters instead of enhancing it. You are only as good as your training.

- **Good communication = good accountability.** We must communicate with our partners verbally and with other members via radio to ensure teamwork and accountability.

- **Physical fitness is essential.** The job of a firefighter is physically demanding, and physical conditioning can save your life. The leading cause of firefighters' death on the fireground is cardiac arrest. Physical conditioning is an essential preventive measure.

Discussion Questions

1. What are some other types of searches that could be used at similar incidents?

2. Other than moving faster than your partner, what are some potential problems when using a TIC?

3. How can a breakdown in communication lead to problems with accountability?

The Forum Restaurant Fire

Retired Battalion Chief
Francis "Pat" Burns
39 Years in the Fire Service
President F. Pat Burns
Investigations, Inc.
Fire and Explosion
Consultant

That cold, gray day, the hawk was howling and the wind chill was well below zero. All of this went unnoticed in the kitchen of Engine Company 104. It was 7:00 a.m., and we were doing our own howling. This day had started out like any other, sitting around drinking coffee and busting each others chops before starting our twenty-four-hour shift. I was a young lieutenant in the Bureau of Fire Investigation, and we shared this house with Engine 104.

A Home Away from Home

In this close-knit firehouse, we were like a family. It was our home every third day. We seldom had a chance to rest in this busy house, so we welcomed the chance to sit around and joke.

Our table consisted of two sections that hinged in the middle so we could fold it to clean the floor. Because the kitchen floor of this old firehouse pitched to a drain in the middle of the floor, the table sloped toward the center. I was sitting in the middle, across from Wally, our engineer. What a character he was. He knew everything a person could want to know about hydraulics—until the bell rang, and then he forgot everything. We learned to keep an eye on him. To my left at the end of the table was Richie. He was about twenty-five years old, had a couple of years on the job, and was a good, tough fireman.

Richie was also a prankster, and I could see out of the corner of my eye that he was trying to pull another one. He had spilled a puddle of coffee on the table in front of him, and I was leaning both arms on the table. Worse, I was wearing a clean white shirt. Winking at Wally, I pretended not to notice the river of coffee flowing slowly toward me.

The liquid wasn't moving fast enough for Richie, and he started flicking his fingers to propel it toward my sleeve. At the last second I raised my arm out of the way as if to scratch my head. Then I turned toward him. "So you want to play, do you?" The full cup of coffee in front of me had cooled a little, and I let him have it, right in the middle of his chest. The guys roared.

Now I had to escape fast. I jumped up onto the table and took off out the door. Richie was right behind me, his cup in hand. I ran up the stairs to my office and made it just in time, locking the door behind me.

Richie pounded on the door, but I ignored him, and eventually he left. Later I found out that he strategically placed buckets of water around the firehouse, awaiting me. Somehow I managed to outfox him all day.

The Alarm Bell Rings

The bells rang about 1:00 a.m. The fire alarm office had pulled a box alarm (a second alarm) on a restaurant called the Forum. The first company on the scene, Engine 25, reported seeing light smoke. The Forum restaurant was a high one-story, bowstring truss building about 75 feet wide by 150 feet deep. There was a second-floor dining loft towards the rear of the building that was served by an inside stairway on both the east and the west sides of the building.

When we arrived, the crew of Engine 25 was leading a 2½″ hoseline through the front door. This line went deep in the building and up the rear stairway to the east. I helped Engine 104 lead a similar line up the west stairway. In those days, our arson investigators assisted the fire companies with extinguishment and would conduct their investigation when the fire was brought under control. As we reached the loft, we could smell burning wood but felt no heat. Only a lazy haze of smoke filled the air.

Only a lazy haze of smoke filled the air.

Our battalion chief, Charlie, was a man we respected for his abilities, but he didn't have the best "people skills." Even though he had a tendency to talk down to people,

With very little smoke in the air, a firefighter opened a small inspection hole in the thick plaster ceiling.

nobody questioned his knowledge. He ordered two truck companies to open the decorative plaster ceiling that concealed the truss area. The truck worked for several minutes trying to open a wire mesh decorative ceiling coated with 4-inch-thick plaster. Finally a small hole penetrated the plaster ceiling. The chief placed a chair on top of a dining table. A firefighter steadied the table as the chief climbed up onto the chair and stuck his head through the hole. After a quick look, he pulled his head out and yelled to a Deputy Chief Fire Marshal who had just come in: "Get everyone out! We lost this building!"

The smoke and heat conditions still weren't that bad. I started to help the firefighters from one of the engine companies pick up their hoseline, until Charlie yelled. "Leave those lines and get out now!"

He had an urgency in his voice and everyone started running down the stairs toward the front door, and I followed them. I was behind the Deputy Chief, and as he reached the front doorway, I was just a step behind him and his driver was a couple of steps behind me. "BOOOM!" It sounded like the world was falling. The entire roof

> *It sounded like the world was falling.*

The chief ordered everyone out of the building—fortunately, as this picture was taken 30 seconds before the collapse.

collapsed at once. When the roof dropped, the pressure blew the chief and me out of the door and into the street. The chief's driver never made it. He was buried under tons of rubble.

I landed in the street on top of the Deputy Chief, and firefighters were scattered on the ground everywhere. Looking back at the building, I observed that most of the roof had collapsed but the front wall was still intact. The large plate glass windows in the center of the building had blown out, but large sections of plate glass were intact at each side of the building. The faces of two firefighters behind the glass on the east side peered out with a panicky deer-in-the-headlights look. Although the collapse has spared them, a secondary collapse was sure to happen soon.

Seeing a blue, wooden street barricade that was left behind by a street repair crew. I picked it up and ran toward the building. Waving to signal the two trapped firefighters to move to the side, I threw it through the large plate glass window. Chunks of glass rained down, and the two men scrambled to safety. I repeated this action on a similar window on the west side of the building and helped pull a firefighter from the rubble.

"Where is 104?" I thought as I ran around the back of the building in search of my fellow firefighters. I knew they were in there, and I had to find them.

Seeing that the building had collapsed on top of engine 104's hose-line, one of the chiefs was ordering Wally to shut his pumps down. Wally refused, shouting, "No—I won't shut down! My guys are in there and they might need this water." He paced back and forth protecting his pump panel like an angry bulldog. Now at the rear, I joined a group of rattled firefighters who were fighting through the rubble to get inside the building. Once inside, we found that a large section of the loft hadn't collapsed but was buried under the collapsed roof. Fire and thick, black smoke were everywhere. The stairway was buried under tons of debris. We had to find a way up to that loft. Our brothers were up there. Frantic, we crawled and dug our way through the debris. Beams of light from our flashlights danced around the ceiling as we looked for access.

> *Beams of light from our flashlights danced around the ceiling as we looked for access.*

"There's an opening!" someone yelled. It was a small opening for a dumbwaiter to send trays from the kitchen to the loft. "We need someone small to fit through that hole."

"I can fit," came a voice through the smoke. A tall, lanky firefighter appeared and stripped off his bulky fire coat. Armed only with a flashlight, he climbed the dumbwaiter, squeezed through the hole, and within seconds disappeared into the smoke. We could hear him moving debris and calling out for his brothers.

Another firefighter clung to the dumbwaiter with his head inside the hole. As we waited, debris fell on us from the unstable structure. Several minutes passed when finally the firefighter on the dumbwaiter yelled, "He's got one!"

Quickly we formed a human chain up the dumbwaiter. Soon a pair of feet in dirty white socks popped through the hole, and we saw the feet move. "He's alive!"

We cheered while they passed him down to a group of waiting firefighters, including myself. We ran him out to a waiting ambulance. The firefighter was Red, the lieutenant of Engine 104. As we laid him on a stretcher, he looked up at me and whispered through his delirium, "They're up there, Pat!"

The next guy they passed down was one of our firefighters Marty. Steam rose off his body as we carried him down the alley to a waiting

After the collapse, several firefighters attempted to bridge a ladder to the collapsed roof from an adjoining building. Rescue attempts where being made simultaneously in several areas.

ambulance. He would tell us later that he had given up hope of making it out alive. When he was found, he was on his knees making peace with his maker.

My buddy Richie didn't make it. We found him later under a beam not far from where the others were rescued. We had to breach the wall on the second floor of a tavern next door to get to his body. I will never forget what he looked like lying there. After thirty-one years, not a week goes by that I don't think about him—how we had goofed around all that day and that night we lost him. Later we also found a third firefighter from Engine 25 who didn't make it.

The lieutenant of engine 104 later told us that when the building collapsed, they crawled away from the heat until they felt a whiff of fresh air coming from somewhere. Then they dug in and shielded themselves with the hoseline. He said, "Thank God Wally didn't shut off our water or we would have been burned alive." Wally had come through when it counted.

Three brave firefighters gave their lives in the line of duty that day— heroes all. God bless them, and may they never be forgotten.

Lessons Learned

- **Bowstring trusses can be misleading.** Because of their height and large bow area, they often hold the heat and smoke high above the firefighter's head. We might have good visibility and low heat conditions at floor level, which gives us a false sense of security. We must realize that the trusses are being exposed to temperatures in excess of 1300 degrees and their integrity is being compromised.

- **Bowstring trusses collapse with no warning.** The only warning sign we can expect is that the trusses are being exposed to a large volume of fire.

- **We should expect a large-area collapse.** Trusses often are placed 20 feet on center. In a 75-foot-wide structure, if only one truss were to fail, we would have a 75-foot by 40-foot collapse area.

- **Evacuation orders must be followed immediately.** When given an order to evacuate, we should not question the order. Sometimes we get tunnel vision and fail to realize that someone sees something that we don't. The rule is to always err on the side of safety.

- **We have to be on the alert for secondary collapse.** Most firefighters die from secondary collapses. After the initial collapse weakens the structure, the impact load often triggers a secondary collapse. Because this impact load can consist of firefighters working, we have to know what is above and below us at all times.

- **The water supply of working firefighters must not be shut down.** If the flow of water has to be stopped, someone must warn the interior firefighters about the situation and wait until they back out of the building.

- **We must remember the basics:** Never give up. Call for help. Stay low, think, and go. Stay calm in bad situations. Never panic.

- **We must learn to react,** to fall back on our training and trust our instincts.

Discussion Questions

1. What do you think the chief saw when he looked through that hole in the ceiling?

2. Obtain the procedure for emergency evacuation from the fire department in your area. What is the basic procedure?

3. What conditions led to the collapse, and what shielded the fire-fighters from these warning signs?

To the Rescue

Firefighter Ed Loder
35 Years in the Fire Service

Most of my thirty-five years of service have been with Rescue Squad 1. In looking back, some incidents leave us with good memories, and some remain sad and painful even after all those years. Regardless of the outcome, we take away a lesson from each.

Rooftop Rescue

It had been a beautiful spring day. Now 9:30 at night, the temperature was still around 70 degrees, warmer than average for late May. We got a still alarm—an alarm called in by phone. A woman was threatening to jump from the top of an eighteen-story hotel. I was the driver of the squad that night, and we all looked up as soon as I started down the block.

At night it was hard to see the upper levels of the hotel. The Haz-Mat unit pulled up behind us. In our town, the Haz-Mat unit carries a portable jump bag that can be inflated quickly, but it is effective only for jumps from about four or five stories. The Haz-Mat unit also carries binoculars, and I grabbed them to scan the top of the building. There on the top floor was a woman sitting on a windowsill, her legs dangling over the side.

"This looks like the real thing," I thought as I tossed the binoculars back into the truck. It was time to earn our pay.

Our lieutenant instructed our crew, "Okay, let's grab our ropes and rescue harness."

I reached for a rescue harness that we called the "bumblebee suit." It was a yellow and black three-point rescue harness. We also grabbed 100 feet of kern-mantle rope and some padding for the roof edge. Our group headed inside, boarded an elevator, and exited on the top floor, where we were greeted by a couple of police officers. The police respond routinely to these situations with a trained negotiator to try to convince the person not to jump.

Soon the negotiator came out to brief us. He said the woman was a psychiatric patient and he couldn't talk her out of it. "She's getting ready to jump," he said.

The police told us they were trying to contact her doctor but things weren't looking good. We walked over to the door of her room and peeked in. She was sitting so far over the ledge that a gust of wind would knock her off. My lieutenant looked at me and said, "Let's go up on the roof and see what we can do."

We raced up a flight of stairs and gained access to the roof through a bulkhead door. We then crept to the edge and peeked over. An ornamental overhang extended about 3 feet from the outside wall. Lying on the overhang, we could see that the woman was barely hanging onto the sill. She was looking into the room and had no idea that we were watching from above. Our District Chief (equivalent to a Battalion Chief) came up onto the roof and said the police told him we would have to do something soon or we would lose her.

Our plan was to set up a rope and prepare to lower one guy over the edge. Our lieutenant asked, "Okay—who's going over?"

"I've got the suit," I replied. "I don't mind going. . . . Plus I'm the lightest one here. You guys are porkers."

They let that insult pass, and our team quickly set up the rope as I slid into the bumblebee suit.

Dope on a Rope

We waited for the command from my officer. He was in contact with the District Chief, who was downstairs with the police. The plan was set: The police were going to distract her, and on a signal I would quickly drop to her level and prevent the woman from jumping. The rope was tied off so I wouldn't fall past her window. I stood on the edge of the roof above her, ready to hop backward over the edge at the Chief's command.

One of my old buddies from Ladder 17 came over and said, "Are you sure you want to do this? You're on the roof—eighteen stories up!" He added, "Any last requests?"

I ignored him.

The lieutenant came over with a serious look on his face. "They just told me she has a straight razor in her hand." He assured me he wasn't kidding.

"Let's get this straight," I said. "I'll be eighteen stories above ground on a thin rope, wrestling a woman with a straight razor."

Everyone laughed.

As we got closer to show time, the mood became more serious. While waiting for the command to go, I stood on the edge of the roof, thoughts zooming through my head: "Is she going to jump at me with the razor?" "Will she come at my face?" "Will she come at the rope?"

> *I stood on the edge of the roof, thoughts zooming through my head.*

It was game time. I stood there ready to go, and my team stood ready to drop me to her level. Sometimes we face challenges on this job that Hollywood couldn't dream up. All I could do at this point was to stay ready and try to mentally prepare for whatever might happen. The Chief's voice came over the radio: "Now!"

Without hesitation, I hopped off the roof. The stop was sudden. The rope stretched like a spring, bouncing me up and down while I tried to keep my eyes on the woman. She was facing into the room as I dropped, but she must have seen me out of the corner of her eye, and she screamed. When she turned toward me, I kicked her with what could have produced a game-winning field goal at a football game. She flew back off that ledge into the room, and a couple of police officers quickly jumped on her.

Still bouncing on the rope like an oversized yo-yo, my heart was in my chest. "Hey don't forget about me," I yelled through the window, "I don't like being out here." I was dangling about 3 feet out from the building, eighteen stories above the ground.

After what seemed like hours, two firefighters came to the window and pulled me in. It felt good to get my feet back on a solid surface. The woman turned and called me a nasty name as she was being escorted out of the room in handcuffs. Mission accomplished.

> *Still bouncing on the rope like an oversized yo-yo, my heart was in my chest.*

"Spider Ed" in action. Photo courtesy of Billy Noonan.

The next morning our Fire Commissioner called to tell me that I did a nice job, and that the guys had tagged me "dope on a rope." Unfortunately, that name stuck for quite a while.

Mayday! I'm Trapped!

A few years later, we were returning from a fire when the call came in shortly after midnight. The dispatch struck a box alarm—which meant that Rescue Squad 1 would be responding along with three engines, two trucks, and—because it was on the water—two fireboats. I was driving, and as we crossed the bridge, I looked down at what seemed to be a small fire on one side of the pier. "We'll have this one out in no time," I told myself. "I might even be able to get some rest tonight."

A large warehouse blocked our path to the fire, which was on the waterfront side of the pier. The engine company already had a line stretched into the warehouse. We followed the line into the smoke-filled building and fumbled around, looking for an exit door that would lead to the waterfront side of the pier. The engines line wasn't charged yet because we hadn't found the door to the fire.

The fireboat's water cannons in action at the warehouse. Photo courtesy of Billy Noonan.

Suddenly the engines of the fireboat started to roar as they charged their big water cannons. We could hear the fire streams hit the rear wall of the warehouse, and within a few minutes heavy black smoke drove us down to the floor. I remember seeing smoke push up from the tongue-and-groove floor and thinking, "This isn't good. The fireboat may have pushed the fire under the pier."

"We'd better get out of here," someone yelled. The conditions changed from good to bad so fast that everyone was disoriented. The visibility was zero, and I couldn't hear anybody. I knew the engine was off to my right, so I started that way. I couldn't find anybody, and it seemed like I was in there by myself. If I could find their line, I could follow it out. Now fire was now coming up through the floor. I found the hoseline and started to follow it, but it was looped around and I didn't know which way I was going. Finally I felt the large lugs of a male coupling, this told me that I was heading the wrong way. I turned around and started back the other direction.

Then I heard my partner, Mike, calling for me, and I followed his voice to the door. Outside, I looked back at the warehouse and the fire was

These two photos were taken a few minutes apart, showing how quickly conditions changed. Photos courtesy of Billy Noonan.

everywhere. It had gone from a small pier fire to a large four-alarm fire in about 5 minutes.

Our officer told us to force open a steel roll-up door. Mike and I were in the process of cutting it open with our saw when the Deputy Chief called for us to report to the other side of the building. He said two firefighters were missing and he wanted us to lead the search for them.

We each grabbed a 1-hour bottle for our SCBA (Self Contained Breathing Apparatus) and some search cables—thin diameter cables 25 feet long with clips at each end. Four of us from Rescue Squad 1 clipped our cables together. Heavy black smoke was rolling out of the door all the way down to floor level, and one of the engine officers stopped us. "Nobody's going in there."

We looked at him and said, "Our brother is in there."

Low to the ground, Mike and I started in the door. Inside, we started to sweep from side to side, searching as we went. While we were pushing forward through the heat, we heard a commotion on the radio. A ladder company had found one of the missing firefighters in a separate area of the warehouse, and a helicopter airlifted him to a local hospital.

Another firefighter was still missing, so we forged ahead deeper into the building until the fire, now burning beneath the pier, got behind us. The area by the door started lighting up, and we had to retreat. Because we didn't have a hoseline to clear a path, we had to dive through the flames to escape. By now the entire pier was afire. Over the radio we heard a faint message: "Mayday! I'm trapped! I'm running out of air!"

"Mayday! I'm trapped!"

Taking a charged 2½″ line with us, we tried to inch forward, but the heat was too intense and our line didn't have any effect on the fire. "Nobody could live through this," I thought.

The fire pushed us out of the inferno for the last time. Our only hope was that maybe the firefighter had found his way out and was below the pier clinging to a pylon. We checked every inch of the pier below but didn't find him.

A couple hundred firefighters used large master streams and dozens of hoselines to try to extinguish the blaze. Throughout the night most of the roof and several of the walls collapsed and we feared that the entire pier would fall into the sea. Finally, around 6:00 a.m., with the fire almost extinguished, the Fire Commissioner came looking for a few volunteers to recover the missing firefighter. He warned us that he wouldn't send more than two, as he wanted to limit potential losses. We knew what that meant. The Commissioner was clear that nobody would be asked to risk their life, but another firefighter, Al, and I volunteered.

Some guys tied a rope to each of us. As I crawled back into the inferno, I looked back at my brothers. By the look on their faces, they would have joined me if they could.

Secondary collapses were still happening around us, and at each collapse, the pier shook. I had a bad feeling, but I had known the missing firefighter for a number of years and was sure that if the situation were reversed, he would come for me.

Slowly and methodically Al and I searched the areas where the firefighter was seen last. We searched about 10 feet apart for close to an hour, sweeping our hands under every piece of debris. Suddenly Al yelled from the other side of a pile of timber, "I found him! He's over here!"

Within seconds, the two of us were kneeling over the body of our missing brother. The microphone of his radio was still in his hand.

I radioed the Commissioner, "We found him, boss. Do you want us to bring him out?" He told us to leave him in place, and we guided members of the lost brother's company to our location. Following the tradition of the Department, members of his own company would bring him out.

When the members of his ladder company arrived, we all knelt and said a few prayers. Then they placed him into a body bag and they put his helmet on top of the bag. Each of us grabbed one of the six strap handles and we carried our brother out of the ruins. When we came out, we faced a sea of blue, as more than a hundred firefighters were waiting. Two lines of dirty-faced firefighters, helmets at their feet, stood at attention and saluted as they formed a path for us.

When we came out, we faced a sea of blue.

This is hard for me to talk about even after all these years. To firefighters, nothing is worse than losing one of our own. The only thing we can do is to try to learn from the experience. Then we shake it off and go to work the next day.

Lessons Learned

◉ **Backup safety lines must be in place.** The rope rescue happened in 1990, and Rescue Squad 1 did an excellent job. Today the fire service has better equipment and procedures. The use of secondary backup lines is now mandatory. One thing that hasn't changed is that it still takes a tremendous amount of courage to go over the edge.

◉ **Vertical rescue is high-risk and low-volume.** High-risk incidents that we do not respond to often are among the most dangerous to firefighters. To increase safety, we must concentrate our training on these seldom-used skills.

◉ **Conditions can rapidly change for the worse—be prepared.** We must always know where we are within a structure in case we have to get out fast.

◉ **Accountability is essential.** Officers must know where their firefighters are at all times. Freelancing is a dangerous practice and is never acceptable.

◉ **Offensive and defensive tactics must not be used at the same time.** Fire attacks must be coordinated so that outside master streams are never used on a building while firefighters are performing an interior offensive attack.

Discussion Questions

1. Using today's equipment, training, and procedures, what would be some considerations at a similar vertical rescue incident?

2. What new equipment and procedures could be used in an incident similar to the fire at the pier?

3. How would RIT or FAST companies be able to assist at the warehouse incident?

Good Smoke?

Battalion Chief
Steve Chikerotis
27 Years in the Fire Service

This must be paradise! I lay suspended between two beautiful palm trees. The sun caressed me and the ocean breeze gently swung my hammock in perfect rhythm with the surf. In the distance, Diamond Head was framed by the bluest sky I had ever seen. Suddenly an elbow struck my ribs. "Hey, Steve—you aren't sleeping, are you?"

Reality Sets In

When I opened my eyes, I found myself nose to nose with the smoke-blackened face of my best buddy, José. It was February of 1981, it was 2:00 a.m., and I was in the back of Rescue Squad 2 returning from our fifth fire of the day, far from that hammock on Waikiki beach. I was tired, wet, dirty, cold, and coughing from the several fires we had fought that day. Yet there was no place I would rather be, with my best friends surrounding me—life was good.

John, who was driving, and our Captain, Bill Burns, sat up front, and two other firefighters were across from me, half asleep. I sat next to José, a bundle of energy who never ran down.

"You amaze me." José said. "How can you sleep sitting up like that?"

I replied, "You amaze *me*. Don't you ever stop?"

"Hey, we did good on that roof, didn't we?"

"Yeah, José, we did good."

He kept going. "You think you're good with an axe, don't you?"

"Yeah, I'm good with an axe," I said wearily, closing my eyes and trying to return to my beach fantasy.

"You think you're better with an axe than I am, don't you?" José persisted.

"Yeah, Joe, I know I'm better than you."

Relentless, José challenged me. "Why don't we have a contest to see who's better?"

Just to get him to stop talking, I replied, "Okay, José, we'll have a contest." Finally I was able to return to my beach fantasy for the short ride home.

I woke up as the rig stopped in front of our quarters. The cold air stung my face as I jumped out into lightly falling snow. Captain Burns went off to write the reports, and the rest of us started cleaning our tools and equipment. One by one, the squad members retreated to the bunkroom. The firehouse was unusually dark and quiet, and the only light on in the house was directly over our squad bay.

Our squad company shared a huge firehouse with two engine companies, a Battalion Chief, a Deputy District Chief, and an ambulance company. Twenty-two firefighters worked each shift, and we were "family." As I finished changing out my air bottle, all I could think about was getting into dry clothes and jumping into my nice warm bed.

The Challenge

Oh, no! Just when I thought it was safe, he was back. "All set!" José announced, proudly holding both of our axes.

He had been busy. A heavy wooden pallet, with a thick white chalkline drawn across it, was set behind our squad. José pointed at it. "We'll see who can come closest to the chalkline."

I replied, "Joe, I can hit a chalkline, but it's 2:30 in the morning. Let's go to bed."

He wouldn't quit. "We'll see who comes closest in simulated smoke conditions. We'll wear blindfolds. Do you want to go first, or me?"

I shook my head in disbelief. "Blindfold me," I said, reaching for my axe. I lined myself up, and José wrapped the blindfold around my eyes. I knew I'd hit the line, and then I finally could get to bed. I raised the axe and took a swing.

> *Suddenly the quiet exploded into the hysterical laughter of several firefighters.*

Suddenly the quiet exploded into the hysterical laughter of several firefighters. I didn't have to tear off the blindfold to realize I'd been set up. Where did they all come from? As if by magic, I was surrounded by everyone in the firehouse except José. He was sprinting away like an Olympic track star.

I stood there shaking my head, afraid to look down. My axe blade was exactly on target, but it had passed through a firefighter's dress cap first, pinning it to the pallet. I pulled up the bill of the cap to read the badge number, 3722. Just as I expected, it was mine. I couldn't believe it, I had just buried an axe through my own dress uniform cap. Before I could race to find José, the bells rang again. I muttered, "Here we go, back in the saddle again."

As we readied our gear, the guys chuckled. They told me that José had been setting up this prank throughout the day and was waiting for the right opportunity to get me. During coffee at shift change, I would be at the center of all the jokes, but the bell thwarted that and now we were heading to another fire. I didn't even have a chance to change into some dry clothes.

One More Fire

I read the building as we arrived. It was a large three-story taxpayer, consisting of ground floor storefronts with two floors of apartments above—circa the 1930s. The stores were closed and secured with burglar gates, the six apartments above them were occupied, and the building was ordinary construction (masonry and wood joist). Flames were blowing out of two side windows of a furniture store on the north side of the building. Heavy smoke was pouring out of the windows on the second floor, and it was starting to light up.

Captain Burns ordered José and me to cut open the burglar gate on the storefront to the north, and then to perform a primary search on the third floor over the fire. Off we went with our saw, axes, and pike poles.

José clamped down on the first padlock with our vise grip and chain, and with two pulls I started the K-12 saw. He pulled back on the chain to

José and Steve worked on the burglar gates. Photo courtesy of the Chicago Fire Department.

keep the lock from moving, and I cut through the shackles in seconds as sparks danced around us. Other firefighters poked through the burglar gates at the plate glass windows with their pike poles, and soon we were shrouded in smoke. José and I repeated this action a half a dozen times, and the storefront gates

> *I cut through the shackles in seconds as sparks danced around us.*

opened. We know each other's moves so well that the whole job took only a couple of minutes. We quickly flipped the saw back into its compartment on the squad, grabbed our axes and poles, and entered the building.

When we reached the second floor landing, we ducked our heads and pulled up our masks. The engine company was inching into an apartment that looked like a blast furnace. The smoke was so thick that you could cut it with a knife, and we still had one more floor to go. Each step we climbed brought a noticeable increase in temperature. As we inched our way up the last set of stairs, the heat was unbearable. The second engine company could be heard leading a line up to our position to back us up.

> *Breaking glass from the skylight started to fall on our heads.*

Breaking glass from the skylight started to fall on our heads. "Nice job, truck," I silently thanked the guys as the heat conditions lessened considerably. Now, with more livable conditions, José and I continued up to the third floor. Upon reaching the third floor landing, we tried to open the south apartment door. It was locked. He yelled, "Fire Department!" and pushed through the door. Once it was open, he quickly shut the door. Now it would serve as an escape route if needed. I wrapped a strap around the doorknob of the north apartment door and pushed it in. The strap would allow me to control the door and pull it closed if conditions were too bad on the other side.

"No windows on the sector-four side," I reminded José.

"No smoke detectors," he shouted in a voice muffled by his mask.

The visibility was zero, as thick black smoke engulfed us and high heat conditions guaranteed a tough, dangerous search. On our knees in the doorway, I reached back and tapped my partner twice on the boot. We commonly used a signal of taps. One meant stop, two—go, and three—back up. "Left hand!" I shouted inside my mask.

José tapped my boot twice, and I left my pike pole at the door. We slid into the front room quickly. Starting our search to the left would be the quickest route to the bedrooms and the windows. My left hand was on the wall, and my right hand swung my axe handle out in front like a blind person's cane. José's left hand was touching my back, and

> *Having danced this dance a thousand times before, we knew each other's moves.*

I heard his axe handle sweeping out toward the center of the room. Having danced this dance a thousand times before, we knew each other's moves. I set a fast pace. Our goal was to get through the living room fast. At this time of night, the bedrooms were our primary targets.

The room we were in contained much more furniture than usual, and the layout was confusing, so it took much longer than we expected to get through what we determined to be the living room and dining room. Finally we reached the first bedroom, and the door was open. By now, the increasing heat was banking us down, and we crawled low into the bedroom. Once in the room, José went right and I went left. Across the room through the darkness, I could hear him exhaling. Reaching the outside wall, I swept my hand across the wall until . . . "Window!" I yelled. I raised up on one knee and with a couple of tugs, the curtains and blinds were history.

José yelled, "Bed!"

I cleaned out the window and the frame with the butt end of my axe while he checked the bed and closet. The heat was increasing. The smoke was pressurized and moving fast out of the window.

"We've got fire on this floor," I shouted, and we continued our search.

Out of Air

By the time we reached the next bedroom, a bell was ringing. I reached back and felt my bell for vibrations. It reminded me that I had 5 minutes of air left.

"Let's finish this room before we leave," I yelled, knowing that nobody could hear me over the bell. Again I found the window, and my partner found the bed. His bell was ringing now, and it was getting hotter. By the time we exited the bedroom, the heat had us banked down to a belly crawl. I took a breath, and the mask clung to my face. I tore off my mask and shouted, "I'm out of air!"

Choking on the thick smoke, I kicked the bedroom door shut and screamed, "Window!" I belly-crawled toward the window, trying to suck fresh air off the floorboards. At the outside wall I stayed low but swept my hand high, looking for the window. I grabbed the sill and threw my head out the window. Flames and thick smoke were lapping up from the first and second floor windows. Gagging and choking, I tried to call out for help. At that time only the chief officers had hand radios in our town. They weren't issued to fire companies until six months after this incident.

My mouth back on the floor, I belly-crawled next to José. "Stairs!" I choked out. He tapped me twice, and we crawled out of the bedroom.

I decided that we must be close to the back door, but the heat seemed to be coming from the back. I turned right to backtrack to the front stairs.

> ### The blistering heat radiated down on us.

José was alongside me as we passed the first bedroom door. The ceiling was lighting up now, and the blistering heat radiated down on us.

"We'll burn up if we rise to windowsill level," I thought. "We have to make the door." We each grabbed the other's coat as we fought to survive.

Every gasp for air brought a gag reflex. Dizzy and weak, I was too tired to go on, but I knew if I were to stop, it would be all over. Choking and gagging, I felt like my heart would explode. Panic tried to set in.

> ### Panic tried to set in.

I told myself: "Keep moving! I'm so hot that I feel like I'm going to ignite, but I'm not going to quit. If my best buddy and I die here, it won't be without a fight. Our families are home now, oblivious to the fact that we are near death. For them, we'll fight until our last breath."

My right hand slid along the wall, and I bumped into what felt like a living room chair. I was weakened to the point that pushing the chair out of our way was like moving a mountain. José and I pushed with all our might and were able to get by it. Still flat on our bellies, we stayed close to the wall and slithered forward.

Suddenly I felt something with my right hand—a glass panel just inches off the floor. A door! I tapped José and rolled onto my side. When I kicked at the door, my foot broke out a panel. Cheek to cheek, my partner and I pushed our faces up against the broken glass and sucked cold air into our lungs. I'll never forget how that air felt—a gift from God, the gift of life.

We were still in a bad spot, but now I knew we were going to make it. The chaplain wouldn't be knocking on our families' doors—not tonight. Flames rolled across the ceiling and 1300-degree temperatures banked halfway to the floor. Even though flashover would soon happen, I felt nothing but exhilaration.

I rolled onto my back and, with José, kicked at the bottom of the French door. It sprang back and resisted for a few kicks before the lock finally broke and the door swung open. We crawled out onto a concrete balcony and pushed the door shut behind us. For what seemed like several minutes, we lay there sucking in fresh air and gagging. I could hear an engine company beating down the flames with the hoseline. "The cavalry has come!" I could feel my heart and head pounding, still too weak to even move a muscle. My eyes were burning and my vision was blurred as I looked over to where José lay. He started to move. I stared as he rolled onto his back, his chest rising and falling rapidly. Slowly his hand undid a buckle and reached into his coat. What was he doing?

What I saw next amazed me then, and still does to this day. He pulled out a pack of cigarettes. I watched in disbelief as he tapped the pack until a cigarette fell out. He put it in his mouth and lit up. I would have been less amazed had he pulled a rabbit out of his hat. I wanted to punch him, but I didn't have the strength. Finally I felt strong enough to speak. "What in the world are you doing, Joe?"

He replied with a grin, "This is good smoke, Steve." I guess we have to write that off as some sort of twisted firefighters' logic.

Lessons Learned

- **We must always read the building before entering and identify a secondary means of egress.** We have to know where our egress points are at all times.

- **We must vent as we go.** When searching, we have to vent. Each window becomes a secondary means of egress.

- **When the low-air alert alarm sounds, we should exit with our partners.** We don't want to get caught off guard when our air is low. We anticipate our alarm going off, and exit in pairs.

- **If trapped, the radio is there to call in a Mayday.** Then we should activate our pass alarm, stay low, and go.

Discussion Questions

1. What tools and equipment do we have on today's fireground that would make this a safer scenario?

2. What would the procedure be for a firefighter caught in a similar situation today?

3. What is your fire department's Mayday procedure?

A Close Call

Retired Battalion Chief
Leo Cox
34 Years in the Fire Service

I was blessed with a thirty-four-year career in the fire service and felt fortunate to spend twenty-two of those years as an engine company officer. If you want a position where you can materially affect the course of a fire, consider becoming an engine officer. Nothing is more important to the successful mitigation of a structural fire than the actions of the first engine company on the scene. If the officer of that unit makes a mistake, the domino effect is amazing. When this happens, everything seems to go wrong from that point on.

As the first arriving engine officer, you have to make critical decisions in a moment's time. These decisions often are based on instincts and experience.

A Factory Fire

It was a comfortable evening in September of 1988, with temperatures in the mid-60s. At about 10 o'clock I was about to pass off floor-watch duties to one of my firefighters when the run came in. Within seconds the motivated crews of Engine 44 and Truck 36 were pushing out the doors. We were headed to our third fire of the day, not unusual for us.

Our destination was only a half-mile from the firehouse, and from about two blocks away we could see we had a hit (a working fire). The fire building, an abandoned, two-story factory with a basement, was about 300 feet wide by 125 feet deep, constructed of heavy timber (mill construction). Heavy fire was showing on half of the first floor, and thick, black smoke poured out of the second-floor windows. The building posed another potential problem because an elevated commuter train ran high above the center of the street. Luckily, the wind was blowing to the north, away from the elevated train.

I asked my engineer, to stop even with the center of the building in the middle lane under the El-train. Immediately I had a firefighter open up the rig-mounted deck gun with a 1½″ tip. Within seconds the gun was directing 600 gallons a minute at the heaviest fire, which was on the sector-two (left) side of the first floor. Another firefighter helped the engineer quickly lay a 4-inch supply line to a hydrant just past the building. My remaining firefighter and I led out a 2½″ line with a 1¼″ shut-off pipe into the center door. From that entryway we had to go up seven steps to reach the first floor, where the fire was showing.

Shortly, we were pushing at the fire with our 2½″ line while our deck gun continued flowing in through windows to the left of us. Our engine alone was throwing 900 gallons a minute at the beast. The heat was banking down, and it drove us to the floor as we slowly dragged the heavy line deeper into the fire. Our progress was slow because we had to extinguish offices and storerooms as we worked our way toward the open factory floor. I had a feeling that the tide was starting to turn in our favor.

I had a feeling that the tide was starting to turn in our favor.

The 6th Battalion Chief saw it differently from his vantage point, and his order came over the radio: "Battalion 6 to all companies: Everyone out of the building! Back out now!" He also escalated the alarm and ordered us to set up a defensive attack. I pleaded with him to give us another minute or two inside, but he didn't budge in insisting that we back out immediately. Against my wishes, we followed the order.

Our deck gun continued to operate, and we kept our 2½-inch line flowing from a vantage point outside the collapse zone. Master streams now were coming in from all directions. Truck 36 was operating an aerial pipe in the front, and Squad 2 was in the rear using the snorkel. The fire I thought we were

The fire I thought we were about to win had turned into a water carnival.

about to win had turned into a water carnival. Firefighters aren't happy when a fire turns into a defensive stand.

Pride Versus Judgment

After about fifteen minutes of filling the building with water, I saw a stream of water inside the building traveling east to west. Someone was in the building stealing our fire! I asked the guys around me, "Where is that stream coming from?"

Someone responded, "Squad 2 is in there with a line."

This couldn't happen. No squad company was going to steal *our* fire! "We have too much company pride to allow this," I advised the troops. "We're not going to be upstaged, especially not by a squad company. Their heads are already too big." We had a rivalry with the rescue squads, and I once designed a T-shirt that read, "Friends don't let friends become squad men."

Quickly I grabbed the radio and convinced the Battalion Chief to order all master streams to shut down. A company was already inside, and we were going in. Soon we were on our way back up those seven stairs to the first floor. As we worked our way in, a river of water was running down the stairs. Much of the fire had been extinguished by this time, and we were able to stand up now. Our truck company was alongside us as we disappeared into the first floor. The visibility was still zero because of the steam and the smoke, which was now grayish in color. Streams of hot water rained down on us as we progressed deeper into the building.

Streams of hot water rained down on us as we progressed deeper into the building.

The Collapse

We were now in the center of the large, open factory area and could see a faint glow of fire near the rear. The hoseline we were advancing snagged on something, and we couldn't continue our forward progress. Rapidly I followed the line back toward the door to find the obstruction. It turned out that a wedge holding the front door open had slipped out and the door had closed on our line.

Just as I pushed the door open and repositioned the wedge—"Booom!"—it sounded like the whole building had collapsed. The vibration knocked me to my knees, and my heart beat wildly through my chest. Had my worst nightmare come true? Had I led my crew to their deaths?

Head down, I charged up the stairs. It felt like I was like swimming upstream. Firefighters were running for their lives down the stairs as I ran up.

"Please, God, don't let it be," I breathed.

I followed the line back to where my company had been, and it looked like the Grand Canyon. The roof and second floor had collapsed through the first floor and into the basement. Dropping to the floor, I looked into the chasm for my company, yelling their names. No answer. Our line was draped over the edge of the canyon, and I could hear it still flowing water. Determined to descend to the basement and get my firefighters out, I ran back down the stairs.

The Reunion

As I hit the landing and started to go down the basement stairs, my missing firefighters grabbed me from behind.

"Man that was close. We almost bought it," they said. "We heard a crack, and luckily we bailed out just in time."

I had to call an ambulance for one of them. The hundred-ton collapse had grazed his heel as they started to run, but fortunately he received only minor injuries. It was a miracle that nobody was killed.

This near disaster was due to my pride interfering with my thought process. It turned out that the squad company wasn't even in the fire building. The stream of water that I saw was coming from a vantage point on the other side of a firewall. That night I used up my "get out of jail free" card, and I learned a valuable lesson. From that point on, I was far more safety-conscious.

Being a good officer also means being a good teacher. I always took pride in teaching my firefighters and preparing them to be future officers. From time to time I had my firefighters who were due for promotion ride the front seat while I rode the back step and allowed them to make decisions, because I was in a position to correct them if they were wrong. I taught them to think ahead and anticipate where the fire was going to be, not just react to where it was at the time. I also preached that when we get to a point where we think we know it all, a fire will come along that will humble us.

The fire that night was one of those humbling experiences. The greatest education you will ever receive on this job comes from learning from your mistakes and the mistakes of others. Always strive to increase firefighter safety through education.

Lessons Learned

⚜ **Mistakes can turn out to be a learning experience.** We should seek to learn from our mistakes and the mistakes of others so that we don't repeat them.

⚜ **We have to be big enough to admit we do make mistakes.** It's okay to make an honest mistake as long as we learn from it.

⚜ **We should use our pride in a positive way.** In the fire service, we need to have pride in our job, ourselves, our company, and our department. Pride is what motivates us to be our best. At the same time, we must never let this pride needlessly endanger the lives of our fellow firefighters.

⚜ **We should utilize the following rules of risk management:**

✓ Take an educated risk to save a life.

✓ Cut back on this risk when only property can be saved.

✓ Take no risk where nothing can be saved.

⚜ **Being a good officer means being a good teacher.** That includes pointing out our own mistakes and telling what we learned from them.

⚜ **We must be cautious when switching from defensive to offensive tactics.** Some specific defensive strategies are:

✓ When we shut down master streams to reenter the structure, we must slow down and proceed with caution. The structure has never been weaker and is in an extremely dangerous state. We have flooded the structure with hundreds of thousands of gallons of water, which weighs 8.3 pounds per gallon.

✓ We should wait a few minutes to let some of the water run off, and to allow the building to settle, before we enter again. Collapse of any portion of the building has weakened other areas of the structure as well.

✓ We must understand that secondary collapse is often triggered by the impact load of firefighters working.

Discussion Questions

1. What conditions probably led to the collapse in this story?

2. What warning signs might precede this type of collapse?

3. How does the concept of risk management increase safety on the fireground?

The Chestnut Street High-Rise Fire

Lieutenant James Altman
25 Years in the Fire Service

We were about halfway through our twenty-four-hour hour shift, and everything seemed to be business as usual. It was November 21st, 2001. Thanksgiving was still a few days away, but outside it looked like midwinter.

"This would be a good night to stay in bed," I thought. Just as Murphy's Law would predict, this was not to be. As lieutenant of Rescue Squad 1, a heavy rescue company, I was in charge of a crew of six.

The alarm came in around 10:00 p.m., reported as a high-rise fire on Chestnut. It would be our tenth run of the day, but this is one I will always remember. Our squad moved quickly through the light traffic, and within a few minutes we turned onto Chestnut Street. As we started down the block, I quickly scanned two sectors of the fifty-story condominium building. Nothing was showing in the windows, but it is usually hard to spot a fire in a large high-rise. A large group was congregating in the lobby, and this usually meant that something was going on.

"We might have a working fire here, guys," I informed my crew. Our squad was the third fire company on the scene. Engine 98 was in position to feed the standpipe system. We spotted our rig behind Truck 3, leaving room to pull off their ladders—unlikely at a high-rise fire. Battalion 1 had just arrived, and we quickly gathered our equipment and followed the chief into the lobby.

Two red lights on the panel in the lobby indicated that smoke detectors had activated on the fourteenth floor. The two banks of elevators had been recalled, and the engine and truck headed up to investigate. Their fire investigative team was to exit the elevator three floors below the alarm and use one of the building's two stairways to reach the fourteenth. When we questioned people in the lobby, nobody seemed to have any information, but they were all jumpy about the alarm going off—understandable, as the World Trade Center tragedy had occurred just two months before.

The elevator came back down under the control of one of the fire-fighters from Truck 3. Our squad quickly entered, and were eager to go up. The chief who was standing near the elevator, looked at me. "Wait for the report," he said.

Just then the radio crackled: "Engine 98 to Battalion 1, we have a fire on fourteen and we're leading out." The elevator doors closed, and we started up. I assigned two of my guys to check the floor above, and four of us would help with the attack and perform search and rescue on the fire floor.

The excitement of going into a high-rise fire is unparalleled. Mostly, I think it comes from the unknown. We can't see anything until we get to the floor, so we don't know what's waiting for us. These fires are always challenging because of the number of potential victims, the large area to search, and the large area for the fire to spread. Also, ventilation is at a premium. We can't open a roof like we can in a house, and usually we can't take out a window, so the superheated smoke quickly banks to the floor. Without ventilation, we really feel the heat from the fire and also from the steam created by applying water to the fire. These fires are always extremely hot.

> *The excitement of going into a high-rise fire is unparalleled.*

"Check your masks, guys," I said as we exited the elevator on the eleventh floor. Even though my crew was highly trained, I worried about my company in those situations.

We entered the stairway, hustled up two flights, and stopped on the floor below the fire. Here we could quickly get the lay of the land, which would help us navigate in the expected zero visibility on the fire floor. At the west end was a long center I-shaped hallway with a conventional stair-way, and near the east end a smoke-proof tower. At each end of the long hallway was a short hall that accessed three apartments. One apartment lined up exactly in the center of the long hallway, with one at each end of the short hall. The engine had just hooked up to the standpipe of the smoke-proof tower, and we followed this line up the stairs.

The smoke and heat were still bearable. The smoke was coming from one of the three apartments on the east end. With the naked eye we couldn't see which apartment the smoke was coming from. One quick look through the thermal imaging camera (TIC) confirmed the location. The center door was glowing white, which meant there was a lot of heat on the other side of it. As soon as the engine's 1¾″ line was charged, I turned to Ocho and gave the order: "Pop it."

He inserted the narrow jaw of the rabbit tool 6 inches below the lock and tapped it in with his hammer. Artie then popped the door in with a couple of quick pumps on the hydraulic unit.

As Ocho pushed in on the now-unlocked door, he encountered resistance. Something was blocking the door. It was an unmistakable feel. "We've got someone!" he yelled.

As Ocho and Artie slowly pushed the door open, flames were leaping out the top. From the other end of the hallway, a woman screamed. I sent another firefighter down there to tell her to stay inside the apartment and keep the door shut. Meanwhile, Ocho was trying to pull the victim we had found, a now-unconscious female, around the half-open door. The engine aimed a stream up at the ceiling in an attempt to battle back the flames. I moved alongside Ocho to help with the victim, who was halfway out the door, but her legs were caught behind the door.

> *Boiling water rained down from the ceiling on top of us.*

Boiling water rained down from the ceiling on top of us, and heavy black smoke blanketed us. With the victim wedged in the door, we lost control of the situation. We couldn't shut the door for our protection if we had to.

No Warning

Without warning, there was a blast of fire from floor to ceiling. It knocked me backward to the right of the door, where I landed on top of another of my firefighters, Kenny. Completely engulfed in flames, we

> *The heat seemed to be melting us to the floor.*

were burning up. Under pressure, the fire pushed out of the door like a blowtorch, blocking our path. We were trapped in that short hallway, and the heat seemed to be melting us to the floor. Pinned down by the

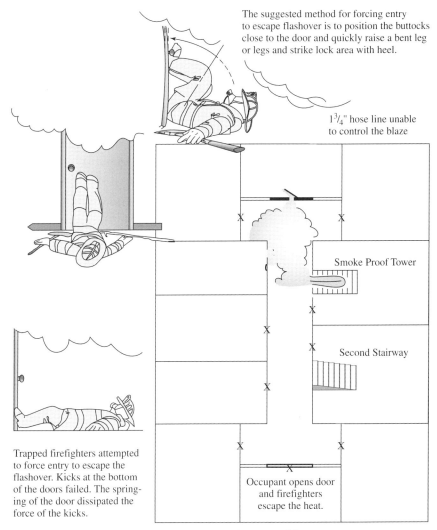

The suggested method for forcing entry to escape flashover is to position the buttocks close to the door and quickly raise a bent leg or legs and strike lock area with heel.

$1\frac{3}{4}$" hose line unable to control the blaze

Smoke Proof Tower

Second Stairway

Trapped firefighters attempted to force entry to escape the flashover. Kicks at the bottom of the doors failed. The springing of the door dissipated the force of the kicks.

Occupant opens door and firefighters escape the heat.

This is a diagram of the hallway and a brief explanation of the action.

intense heat, I tried to force in the door next to the fire apartment. If we could get in there behind the door, we'd be all right, I thought. Lying flat on my back, I kicked at the bottom of the door with all my might, but it didn't budge.

While this was happening to us, Ocho was blown to the left of the door. He tried to grab the victim and pull her out with him. No luck. She was stuck. He saw her skin bubble up, and he finally had to let go. He grabbed the $1\frac{3}{4}$" line, which had been abandoned, and tried to drive the flames back, but the stream had no effect on a fire of this magnitude. Badly burned himself, he had to dive down the stairway.

After failing to force the door next to the fire, Kenny and I were trapped by the blowtorch in a 4-foot-by-6-foot area. I put my face piece up to his and told him, "We have to go or we'll die here."

He got the message and quickly replied, "Let's go."

We dived through the flames. It felt like a million bee stings. We crawled as fast as we could try to get away from the heat.

Out of nowhere, I was kicked in the head, knocking off my helmet and face piece. I lost my helmet and struggled to reposition my face piece as I pressed on. Thinking back, it was probably Ocho's boot that hit me as he dived down the stairway.

Kenny yelled, "My boot is coming off! My leg is burning!"

I shouted back, "Pull it up, and keep moving."

Kenny and I ended up missing both stairways and banging headfirst into the apartment at the opposite end of the long hallway.

We're Burning Up!

Even at the far end of the hall, intense heat had us pinned to the floor. This is the area from where I heard a woman screaming when we started our fire attack. We banged on the bottom of a few doors and yelled, "Fire department! Open up!"

Finally a woman shouted back, "I'm in here!"

"Open the door! We're burning up!" I shouted.

I heard her at the lock, and I knelt up into the heat to reach the doorknob. It was still locked. I fell back to the floor and yelled for her to try again. After what seemed like an eternity, she unlocked the door and we rolled into the frightened woman's apartment. One of the firefighters from Truck 3 dived in with Kenny and me, and we slammed the door behind us. He had been at the far end of the hall trying to find the stairs, and he followed us to safety.

The door proved to be a great barrier. Inside the apartment there was only light smoke conditions and no heat. We opened a few windows to let in some cool air, hoping it would ease the stinging pain. I called Battalion 1 on my portable radio, gave our location, and told him that we were burned. He told us to hold tight and they would get to us as

> *The door proved to be a great barrier.*

soon as possible. I kept trying to reach the rest of my company, but there was too much congestion on the radio. Kenny started to pull off his boot, and I could see that his leg was badly burned. I told him to leave the boot on until after we got out.

Skin was falling off my left hand and my ears, I had taken my glove off to feel for heat conditions just before it lit up, and I tore my hood down trying to put my face piece back on after being kicked. The poor woman in the apartment looked on in horror.

When the pain got unbearable, I said to Kenny "We have to get to the stairs." I asked him if he could make it down the stairs, and he said he could.

The woman didn't want us to leave. I assured her that she would be all right, and we were leaving the other firefighter with her. Kenny and I slipped back into the hallway. Another company was now battling the blaze, and I knew they were committed to putting it out and doing the search and rescue. We found the stairway and hobbled down all the way to the lobby.

I found out that all my firefighters were accounted for. Then I was helped into a waiting ambulance, and Kenny and I were off to the hospital.

The second engine company finally extinguished the fire with a 2½″ line. The woman that Ocho had tried to save was found dead. All of the other occupants, including the lady who opened her door for us, escaped unharmed. Eight firefighters from our initial group were burned and transported to hospitals. Kenny, Ocho, and I were transferred to burn units. Over time, all of us healed and went back to crawling the hallways again.

The Aftermath

It is said that what doesn't kill you makes you stronger. As I lay in a hospital bed, I had a lot of time to think about this incident. I reflected on my family—my wife and kids—and how they would be affected if I didn't come home. I thought about my firefighters, their families, and the woman we almost saved. A single death affects a lot of people.

> *What doesn't kill you makes you stronger.*

The high-rise fire brought us closer together. The whole experience made us more safety-conscious. Now I faithfully check to be sure that all of our firefighters wear the proper protective gear. We train harder, work safer, and vow never again to get caught with our guard down.

Lessons Learned

⚫ **The blowtorch effect that we witnessed was caused by a number of factors:**

1. *High fire load:* The fire was fueled by an abnormal amount of furniture, clothing, and papers.

2. *Strong winds:* The large window in the fire apartment failed due to the heat as the firefighters were trying to pull the victim out of the door. If wind is already strong at ground level, we can expect it to be even stronger on the fourteenth floor.

3. *Stack effect:* The temperature was below zero, and the fire floor was a low floor (fourteenth floor of a fifty-story building). This stack effect caused an inward rush of air as the window failed. In cold weather the lower floors in a building will experience this inward rush of air (the lower twenty stories of a fifty-story building). In cold weather the top floors would have an outward rush of air, which would aid ventilation (the top twenty stories of a fifty-story building). The center floors are a neutral zone where the stack effect is not a factor. An easy way to remember this is: Low/Low is a No-No, or Low temperature and low windows = Danger.

⚫ **The proper-sized line is important.** The initial attack line was a 1¾″ line with an adjustable fog pipe flowing 150 gallons per minute, which had little effect on the fire. The second company used a 2½″ line with a smooth bore nozzle with a 1¼″ tip flowing 320 gallons per minute, and the fire was quickly extinguished. After this fire, the 2½″ line with 1¼″ shut-off pipe became our Department's standard lead-out for a high-rise fire attack.

⚫ **A solid stream should be used in limited ventilation situations.** The solid stream has a greater horizontal reach, better penetration, more gallons per minute, and—most important—promotes less steam conversion, which means less heat for the fire attack team.

Discussion Questions

1. Obtain your fire department's Standard Operating Procedures for high-rise fires. What do these consist of?

2. How do both the larger-diameter hose and smooth-bore nozzle improve firefighters' safety on the fireground?

3. Why is it important to control the door during forcible entry?

Sharing Experiences

Captain Edward Cerdan
26 Years in the Fire Service

In just two hours my wife, Susan and I would be back home. Our vacation had been fun but exhausting. Now her head rested on my shoulder, the gentle motion of the airliner rocking her to sleep. I sat with a notepad and pen, reflecting on an incident I've never told her about.

Guard Duty

It was a sweltering day in July. I was the Captain of Truck 27, housed with Engine 62 and the 22nd Battalion and charged with protecting a southside area of the city.

We had just finished a light lunch after spending the morning cleaning the firehouse, truck, and equipment. We also managed to fit in a ladder-raising drill. In between, we responded to a few medical calls, but we hadn't smelled smoke yet today. I sat at my desk on the apparatus floor, writing up some reports while the sun beat down on me through the window. Outside the temperature was well above 90 degrees, and inside it wasn't much cooler.

Otis, our firehouse dog, kept walking by my desk carrying a board, a two by four about 3 feet long. I wasn't paying much attention to him because he always seemed to have a stick in his mouth. A while later he stood by the open overhead doors barking at two construction workers in orange vests. Otis was a big, black, mixed breed, mostly German Shepherd, and his job was to protect our firehouse from anyone not in a uniform. Those who were outside kept their distance, not knowing that he wouldn't even bite a burglar.

"Otis!" I yelled, and he stopped barking.

"Hey, Chief," one of the construction workers called to me, "we're trying to put in forms for a new sidewalk down the block, but your dog took all of our stakes."

We looked around and, sure enough, in the back of the firehouse we found a pile of about twenty two-by-four stakes. Our four-legged thief had been busy all morning.

Our four-legged thief had been busy all morning.

After a good laugh, a few of the firefighters helped me return the stolen goods. Otis pouted.

The crime solved, I went back to my paperwork. As soon as I started, one of our neighboring companies was dispatched over the fire radio: "Engine 75, report of a rubbish fire on 123rd and Indiana." A minute later an excited voice came across the radio: "Engine 75 to Englewood—we've got a fire. Give us a full still [two engines, two trucks, and a Chief] at this location." I rang the bells, and we sprinted to the rigs. Within seconds Otis was alone in the house, back on guard duty.

As we pulled out of the firehouse, I turned to my crew. "We've got a hit [a working fire]. Engine 75 is already on the scene." From a mile and a half away we could see a thick black plume rising into the sky to the southeast. Engine 62 led the way through the mid-afternoon traffic, and we arrived within two minutes.

The fire was in a one-story bowstring truss battery factory. The building was about 200 feet wide by 125 feet deep, and a warehouse area to the

Otis guarding his loot.

left was fronted by three large overhead doors. To the right of those doors was an entry door that led to both, the offices on the right, and the warehouse on the left. Heavy black smoke was rising from the rear of the warehouse.

Engine 75 already had led a line around the outside to the rear of the building. The intensity of the fire was growing rapidly, so we had no time to waste. I figured if we could open up the overhead doors quickly, we could hit it from the front before the trusses would collapse.

Taking one of my senior men, Mike, with me, we headed for the front door. Other members helped the engine and vented windows from the outside. As we passed the Siamese connection for the sprinkler system, I whistled to the engineer of Engine 62 and pointed out the connection. This could save the office area.

Mike and I forced the front door, and we entered a dark hallway filled with light smoke. A door led to an office area to our right, and another door at the end of the hallway to our left led to the warehouse, I surmised. As Mike forced open the door to the warehouse, I heard the chief on the radio: "Battalion 22 to Englewood. Give me a box alarm . . ." (a second alarm).

The door was no challenge to Mike, and within seconds it was open. Thick gray smoke started to pour into the hallway. Peering into the warehouse, I could see that the gray smoke was banked down to the floor, but there was not a lot of heat. I quickly closed the door so we could mask up and I could relay my plan to Mike: "I'll lead. Left hand search gets us to the front, and we'll find a way to open those overhead doors." We quickly pulled up our boots and put on our masks. "Ready?"

"Let's go," Mike nodded, and I opened the door. He slid his bar in the hinge side and tore the door completely off. I set the door off to the side in the hallway. If the conditions were to get too bad, this building could collapse, and I wanted to be able to get out fast.

We slid into the warehouse, and I kept my left shoulder against the wall. Mike had a hand on my boot as we crawled toward the front. I could feel large tanks up against the wall. "Probably oxygen or acetylene," I thought. We crawled over tools and pieces of steel, which stabbed at our knees as we passed. I felt a wheel to my right. A large truck was almost blocking our path, but we squeezed through. Slipping across the greasy floor, we made our way past the obstacles. The fire was getting worse, and thick black smoke now filled the large warehouse.

"Cap, where are you?" Mike's voice was muffled inside his mask.

"Right in front of you, Mike."

He grabbed my boot, and we pushed on. We encountered even more obstacles, and I still couldn't see the door.

"I don't like this, Mike. Back out and we'll open the doors from outside."

Foiled

As I was turning to follow Mike out of the building, I thought I saw the rail of the overhead door. Too late. I followed Mike back toward the hallway door.

Back outside, Mike went to the truck to get a saw to cut open the doors. I couldn't shake the thought, "We were almost there We should have been able to open it. I know I can get it. I'm going to try one more time." I disappeared into the smoke by myself and turned left into the warehouse, hugging the wall, retracing my steps. I felt the tanks, the steel, the tools. "Good. I know where I am."

When I squeezed through the tight space between the truck and the wall, I started to feel heat. The conditions were changing rapidly for the worse. The black smoke was much thicker, and it was moving fast. "This isn't good," I told myself, "It's going to light up soon." I weighed my odds for a split second and reaffirmed that I had to get out of there—and quickly!

Turning around, I started to slip on the greasy floor as I squeezed past the same obstacles as before. Again I felt the tanks. Good.

It was getting dark, and I couldn't see my hand in front of me. My mind raced. Is this light working? Where am I? Where's the wall? Where's the truck that was parked here? I reached in all directions and felt nothing. I swept out with my leg and still felt nothing.

Conditions were getting worse, and nobody even knew I was here. Reality was hitting me hard. I was lost in a truss building that could collapse at any second.

Conditions were getting worse, and nobody even knew I was here.

Moving forward suddenly. I felt something—a chair. I reached forward and felt a desk.

Now I knew I was lost. How did I get over here? Somehow I must have crossed over into the office area on the right, and now I didn't have a clue which direction to go to take me out. Meanwhile, the conditions were getting worse by the second.

"Stay calm, Ed," I told myself. "Remain calm. You're okay. You still have air."

I stayed still and tried to get my bearings. How much air did I have left? I grabbed my chest gauge and shined my light on it. The smoke was so thick that I couldn't read it. There couldn't be much air left. I grabbed my radio: "Truck 27 to Battalion 22." No answer. My low-pressure alarm went off. That meant that I had only a few minutes of air left.

Was I on the right channel? I took off my glove and counted the clicks, knowing that four clicks meant I was on the fireground. I tried again: "Truck 27 to Battalion 22." Still no answer.

Now I was really scared, and the faces of my wife and two daughters appeared in the blackness. Could this be the end of the line for me? I've heard of people whose lives flashed before them when they were about to die, and now it was happening to me! I fought the urge to panic and tried to think logically. I held my light in front of my face piece, pointing straight up. I knew the smoke would move away from the fire, so I looked in that direction. With only seconds of air left, I had to stay low and go in that direction.

> *Suddenly with my chin to the floor, I saw a ray of sunshine.*

Suddenly with my chin to the floor, I saw a ray of sunshine, just a glimpse, pointing me in the right direction. "There it is—the way out!"

I dived out into the sunshine and breathed, "Thank you, Lord." Throwing off my mask, I looked around. The other firefighters were busy working. Nobody even noticed I was missing! That was because the fire was winning and by now had progressed to a 2-11 alarm (third alarm). We still had a job to do, and I had no time to think about what had happened. I had to pick myself up and get back on the horse. With my company, we opened the building and set up for aerial pipe operations.

Over fifty firefighters were now on the scene. For more than two hours we battled the blaze in extremely hot conditions. Finally, we brought the fire under control and my company entered the unstable remains of the structure to overhaul. As my company opened up sections of collapsed roof looking for small pockets of fire, I retraced my steps to see where I'd been.

One of the mistakes that almost cost me my life was momentarily taking my right hand off the wall on my way back to the hallway door. Leaving the wall for a second as I danced around pieces of scrap steel caused me to miss the door and continue toward the rear of the warehouse. The confusion was magnified when I felt the desk and chair. This caused me to believe that somehow I had crossed the hall into the office section, and now I was totally turned around.

As it turned out, the desk belonged to a warehouse foreman, and I was still in the warehouse only 15 feet past the doorway. My biggest mistake was going in alone, and not letting anybody know I was in there. What saved my life was that I resisted the urge to panic. I stayed low, read the direction of the smoke, and moved in the opposite direction.

When the overhauling was completed, we picked up what seemed to be a mile of hose, as well as our tools and equipment. Finally, it was time to head back. Our tired, dehydrated crew staggered back to the truck. The short ride back to quarters was the first chance I had to reflect on what had just happened. I didn't mention it to anyone that day. I guess I was embarrassed that a captain with twenty-three years on the job almost died fifteen feet from the doorway.

The Critique

The next workday the Deputy District Chief summoned all of the companies that had worked at the 2-11 to our firehouse for a critique. After going over the tactics that were used and a few "atta-boys," the Chief asked if anybody had anything to add. There was the usual ribbing of fire company versus fire company, each one bragging that they had saved the day.

The whole time, I was sitting there debating myself, "Should I say anything? Should I admit I messed up? That's a hard thing to do, but I don't want another firefighter to make the same mistake. Maybe next time the ending won't be the same." I cleared my throat and began.

Confession

When I told my story, the place got quiet. Later, some of the young firefighters came up to me and said that if it could happen to me, it could happen to them. My story had hit home. Sharing our lessons learned when we make a mistake is a way of making the job safer. I learned some valuable lessons at that fire, and I won't get complacent again, not in this job. Even if we do everything right, the unexpected is just around the corner. We have a responsibility to share our experiences with other firefighters.

Aftermath

That day on the airplane, as I was finishing this story, my wife woke up and asked what I was writing. I don't like to worry her, so I hadn't told her about the incident. This time she had me cornered, so I let her read my notes. Her eyes teared up. The date of the incident was July 21, the birthday of her brother who had died years before I met her. She said it was her brother who sent me that ray of sunshine to guide me out. Who knows?

Lessons Learned

- **We always should work in pairs.** Freelancing is too dangerous for the fireground.

- **We can never repeat this advice too often: Remain calm.** Civilians panic, but firefighters must stay calm so we can react to the situation with a cool head. We have to stick with our basic survival skills, call a Mayday, stay low and go, and never give up.

- **Some zero visibility search tactics and tools that could be used are these:**

  1. Left hand or right hand search. (Maintain contact with the wall at all times.)

  2. Thermal imaging cameras (TIC). (Never be completely reliant on the camera, as electronics can fail. Use it to supplement your basic skills.)

  3. Search ropes. (These are especially valuable in large-area searches.)

  4. Leave your partner at the door. (The partner stays at the door or opening and verbally guides you back.)

Discussion Questions

1. Does your fire department have a Mayday procedure? If so, what is it?

2. Does your fire department use RIT (Rapid Intervention Teams) companies at working fires? If so, what are their procedures?

3. How could accountability have been improved at this incident?

A Noble Breed

Firefighter Jimmy Sandas
32 Years in the Fire Service

It was a Saturday afternoon in Brooklyn, and I was at "The Rescue" on Bergen Street. I was a firefighter with the City of New York, assigned to Rescue Company 2, considered to be one of the busiest fire companies in the nation. The firehouse was my second home. My family at "the rescue" consisted of five other firefighters, and an officer.

On this hallowed ground, we always felt the presence of our comrades who made the supreme sacrifice. On September 11, 2001, seven of our brothers hopped aboard Rescue 2 and raced to the World Trade Center, not knowing this would be their final destination. Several more of our brothers, who had been promoted out of Rescue 2 and were working on other fire companies, also gave their lives that morning, including our former captain, a man we believed could walk on water, Chief of Special Operations Raymond Downey. Even though they are gone, we can feel their spirit within these walls.

Three Years Later

The date was January 17, 2004. Outside, the air was cold and the sky was gray. Light snow was falling when we rolled out the doors to fight another fire. Lights were flashing and the siren was wailing as Bobby, our chauffeur, guided us through the narrow streets. Even though it was only 4:30 p.m., dusk was falling as we raced to our third fire on the day tour.

When we pulled up, all eyes sized up the building. Heavy smoke was pouring out of the third floor and attic windows of a 3½-story Queen Anne frame structure. The building had been converted into four apartments, all of which were occupied, and there was a report of people trapped. The Chief transmitted the "All Hands Alarm," indicating that all companies were being put to work. Two hoselines were already stretched into the front stairway. The Chief was in front of the fire building, pointing up at the top floor. He said to our boss, "Lieutenant, can you make the attic and give the guys a hand up there?"

Without hesitating, we split into teams of two and hustled up the stairs, high-stepping over the hoselines and into the darkness. The first line stretched into the third floor, facing a medium fire condition and heavy smoke. One of our teams disappeared into the doorway. Our two remaining teams continued up the stairs, following the second line into the attic. I had a bad feeling about this one. "Something's going to go wrong at this fire," I predicted silently. It was getting hotter every step we climbed until we finally reached the attic.

The second engine company was making a push down the attic apartment hallway. The Rescue's SOP—standard operating procedure—was to split the company into teams. For this fire, one team would follow the engine down the hallway and search for trapped victims. The second team, upon entering the attic, would go to the right and make a primary search of the bedrooms and living quarters.

Our search proved to be negative. We found nobody. The heat was increasing, which indicated a lot of fire somewhere. We started pulling ceilings and opening up the walls, which exposed a medium fire condition.

The engine started to extinguish the fire. The smoke was getting heavier, and the heat was intensifying by the minute. As we continued pulling the ceiling in a front bedroom, a firefighter outside on a portable ladder got our attention: "You guys are about to get pounded!"

I stuck my head out the window and looked toward the other half of the building. Heavy black smoke was pushing out of the windows and lighting up. It looked like the engine was losing the battle. We continued to pull the ceiling when suddenly Bobby said, with urgency in his voice, "Jimmy, we have to go—now!"

When a senior firefighter with more than twenty years experience says, "Let's go," we don't question; we follow. As soon as we started for the entrance to the attic, the ceiling lit up. Thick, black, superheated smoke and intense heat filled the attic, forcing us to craw on the floor. This raging inferno was so fierce that a couple of firefighters escaped from the attic by bailing out of windows onto portable ladders.

I followed my partner, both of us crawling low to the floor, and turned down a hallway toward the attic door. In the zero visibility I could feel a couple of firefighters log-jammed in the doorway. Company policy dictated that our job was to make sure that everyone was able to get out safely, and rescue firefighters had to be the last to leave the fire floor.

> *Suddenly, I heard a man's voice scream out from somewhere deep in the apartment.*

Suddenly, I heard a man's voice scream out from somewhere deep in the apartment: "Help! I'm in here."

Hearing the screams, I stopped in my tracks. My partner was already safely out the door. I had to make a split-second decision. I knew that within seconds this attic would be an inferno, the door to safety was right in front of me, but someone was trapped, and his life depended on me. I had no option. "The game is on."

Following the Voice

Turning toward the sound of the voice, I belly-crawled down the hallway, back into the unknown. The man called out again, and I followed his voice. Maintaining contact with the wall with my left shoulder at all times, I came to a bathroom. There he was—a firefighter was pinned down by the superheated conditions. Now face to face with him, I ordered, "Grab hold of my back."

I went back into the hallway with the firefighter clinging to my Scott Air-Pak. To maintain my bearings, I did the reverse—keeping my right shoulder to the wall. The temperature was getting hotter by the second. Even at floor level it felt like my body was baking inside my bunker gear. I thought to myself, "We're not going to make it. This attic is going to flash any second. It's one of the hottest fires I've been in during all my years as a firefighter. Where's the door? If we don't find it fast, we're going to die here!"

> *All of a sudden a beam of light appeared.*

All of a sudden a beam of light appeared. Not knowing if it was a firefighter's flashlight at the attic entrance or it was the entry to the pearly gates of heaven, we both crawled toward the light. We dived out the door at the moment of total flashover. A couple of firefighters grabbed us and pulled us down the stairs.

We both tore off our face pieces as we staggered down the stairs. On the half-landing we pulled down our hoods and opened our coats to cool our inner core body temperature.

The "All-Hands" Chief assigned to the floor above position radio-transmitted a Mayday because of the flashover and the confusion about possible missing firefighters trapped inside the attic apartment. I heard my lieutenant calling for me on the radio and tried to answer a few times, but others were "stepping on" my message.

Now my lieutenant was calling my name: "Jimmy, where are you?" A tough, muscular guy, he never seemed rattled, but this time I sensed alarm in his voice.

Finally, I was able to get through on the radio: "I'm okay. I'm down here on the second floor landing."

The incident commander immediately conducted a roll call. All members were accounted for.

Shortly, my company surrounded me outside the fire building. Bobby asked me, "Where did you go? Why didn't you come when I called you? I thought you were behind me."

"I was," I replied, "but then I heard a firefighter calling for help, so I went to get him."

Bobby asked, "What firefighter?"

"The one who was lost in the attic." I went on to explain what had happened.

Later, we found out who the firefighter was. He was a probie from a ladder company who had just graduated from the fire academy the week before. This was his first fire. In all of the mayhem, he got separated from his officer and another firefighter. He wasn't burned at all, just shell-shocked. In the confusion, his officer didn't realize that his probie was missing, and when he did, he saw the two of us crawling out of the attic. He thanked our company for pulling out the probie.

"Forget about it—it's our job," I responded. "We're just glad he's okay."

Staring at me, the truck officer said, "You have a nasty burn on your face." I could tell where I was burned because I felt the throbbing and stinging where the mask outlined my face. The probie had accidentally pulled on my regulator hose, which shifted my mask and broke the seal with my protective hood. By the next morning, it looked like I had been attacked by killer bees. My face was swollen, with blisters from an ear down to my mouth, and my left eye was swollen shut. It was time to go the hospital Burn Center.

A Wife's Call

My wife Donna always called the firehouse in the morning to see what kind of night tour (shift) we had. This time she must have had a feeling that something wasn't quite right because she called in the middle of the night. Being the wife of a firefighter isn't easy—always wondering but never saying aloud if her husband will get hurt—or worse—in the back of her mind.

When she called, one of the new firefighters assigned to "The Rescue" answered. Not knowing how to respond, he just said, "Jimmy isn't here."

"What do you mean he's not there?"

"Put Woody on the phone," she said.

Woody, my close friend, got on the phone, and my wife went to work on him: "Woody, I'm getting too old for this. Is Jimmy all right?"

"He's fine. They're just checking him out in the Burn Center." He then explained what happened. Woody then added, "How do you do this? How do you always know?"

His questions went unanswered. A minute later she was on the phone to the Burn Center, and a nurse put me on the phone with her. "Woody told me you rescued a probie. How are you?"

"Remember what you said . . . when something like this happens, think about me and the kids. Well, I was thinking about you and the kids. Our son Michael is going to be a probie someday, and I would want someone to rescue him."

That was all Donna needed to hear. She was also thinking about our eighteen-year-old son, who could hardly wait to follow in his dad's footsteps.

Postscript

Although the realization of the dangers of this job has always been there for the firefighter's family, Jimmy's wife worries more now than she did in the early years. The reality hit hard on June 17, 2001. They lost their best friend, a brother, Brian Fahey of Rescue Company 4, who died with two other firefighters, Harry Ford and John Dowling, on what became known as "The Father's Day Fire" in a neighborhood hardware store.

Then on September 11, not quite three months later, the off-duty fire-fighters from Rescue 2 grabbed their gear, stripped the firehouse of every available piece of equipment, commandeered a bus, and headed to the burning Twin Towers. While they were enroute, the second tower collapsed. Like the other spouses of firefighters who were working there, Jimmy's wife didn't know if he had survived. On that dark day, time passed slowly, and she grew more nervous by the minute. Friends kept calling and asking, "How's Jimmy?"

Finally, sometime after midnight the phone rang. It was Jim calling to say he was all right. On September 11, 2001, 343 heroic firefighters gave their lives, including all seven on-duty firefighters who responded from Rescue 2.

September 12 dawned warm and sunny, but the site was shrouded in smoke and haze. As sweat poured off their bodies, the firefighters, covered in a layer of dust and eyes burning from the contaminated air particles and smoke, continued the search. Every minute that passed without recovering their brothers increased the pain that stabbed at their heart.

Thousands of rescue workers—firefighters, police officers, ironworkers, and heavy equipment operators—worked side by side, yet it was eerily quiet. Suddenly a guy tapped Jimmy on the shoulder and handed him a cell phone. It was his wife: "Jimmy, are you all right? Have you found any survivors?"

Puzzled, he asked her, "How did you find me? Where did you get this firefighter's cell phone number?"

After a long pause, she said, "Take a look at him, Jimmy."

Standing in a pile of debris about five stories high, his face covered with dust, Jimmy looked into the fireman's face. It was his son, Michael. Early that morning he had found some of his father's old turn-out gear in the garage, traveled forty miles, and sneaked through checkpoints to reach the site. Amidst thousands of dust-covered workers in that place, this future firefighter risked his own life, and somehow managed to hunt down his father. Firefighters, even future ones are truly a noble breed.

Lessons Learned

- **Officers have to know where their firefighters are at all times.** If members aren't accounted for in a hazardous situation, they must call a Mayday immediately. There's no time to waste.

- **The warning sign of flashover** is a rapid, intense heat build-up. The lower the ceiling, the quicker the flashover.

- **Proper hose lead-outs are the key.** The single most important tactic to control the fire, and protect firefighters, is a proper lead-out. We must have enough line in the proper size, stretched so it can reach the entire fire area.

- **We are instructed to use a righthand or lefthand search in zero-visibility situations.** Count walls, doors, and windows, and know where you are at all times.

- **Visibility is always better at floor level.** If lost, we must never give up. Instead, call a Mayday, sound the pass alarm, stay low, and go.

- **Risk management is necessary for firefighter safety.** Training is the key. As we say, we sweat in training so we don't bleed in battle.

- **Thinking about our families gives us the incentive and motivation to carry on.**

Discussion Questions

1. How could a RIT or FAST company have assisted in this rescue?

2. What are some other pieces of equipment that could have aided in this rescue?

3. What are some dangerous situations that could arise at a similar incident, and what are their warning signs?

A Case of Arson?

Battalion Chief Steve
Chikerotis
27 Years in the Fire Service

On that cold winter day in 1989 I was a relief lieutenant detailed to Truck 13 for the day. At that time in my career I had a route of five fire companies where I would serve as the company officer. Truck 13, which was housed with Engine 106, was one of my favorite details. It seemed like we always caught fires when I worked there, and these were two good fire companies. Everyone in this house was hard-working and dedicated, and when there was work to be done everyone pitched in. Besides, the firefighters in this house were among the funniest characters I'd ever met. Someone was always the brunt of a prank. Today was no exception. We joked around all morning, with the exception of one fire around noon. The rest of the day was quiet.

Restaurant on Fire

I was sleeping like a baby when the next fire run came in. It was 3:00 a.m. We were the designated still alarm (first alarm) truck to a restaurant fire. Within seconds the troops were sliding down the pole and running to the rig. These firefighters took great pride in their push-out—another reason I liked working at this house. The Battalion Chief arrived at the fire ahead of us, and his voice sounded excited on the radio. This caught our attention because he was not an excitable guy. We were about to find out why.

We read the building on arrival—a one-story, greasy-spoon restaurant, about 40 feet wide by 80 feet deep. Flames from front to back were showing through the large front windows. The business was closed for the night, and the front glass door was locked. The engine company was starting to lead out a 1¾-inch hoseline, and I was deploying my troops: two firefighters to the roof, one to open the rear, and one with me to force entry in the front and start to search.

The chief called me over. "Steve," he said, "that newspaper driver over there was stopped at the red light in front of the building. He swears he saw a line of fire swish across the floor and totally engulf someone. He said that person was running around inside while on fire."

"We'll get him, boss." I grabbed my radio, and we hustled towards the door. "Truck 13 to truck 13 rear. Bobby, there's supposed to be someone in here. Let me know right away if the rear door is already opened."

"Take out the front windows, Paul," I said as I rushed toward the front door. One swing of my halligan bar took out the glass in the steel-framed door. With two more swings, the cross-bar handle was knocked free for easy entry. An officer and a firefighter from the engine company climbed through the doorframe and pushed forward with their 1¾-inch hoseline. Glass from the front window could be heard crashing to the ground as the engine company crouched low and disappeared into the black smoke.

> ## I could smell the strong odor of gasoline.

As I pulled my mask on, I could smell the strong odor of gasoline. I slid in behind the engine Paul, who had taken out the two large windows, masked up and followed me in. Holding my radio close to my mask, I relayed, "Truck 13 to Battalion 10—strong gasoline smell in here, boss."

The response came: "Truck 13 rear to Truck 13. The rear is locked tight. I'm forcing entry."

"Message received," I replied, then said to my partner. "It's arson, Paul, and he's still in here."

The engine was battling a stubborn fire as we swept out and searched. Paul checked the dining area and I dived over the counter and checked the grill area. There was no sign of the arsonist as I worked my way to the rear. When I met up with Paul, he reported the same result.

The fire was finally being brought under control. I gave the chief an update, "Truck 13 to Battalion 10, primary search is complete and negative. Secondary search is under way." I told myself: "He's got to be in here. Where is he?"

In a rear area behind a bathroom, I found a doorway leading to the basement stairs. "Let's go," I said to my partner. Quickly we descended into the dark basement. The fire hadn't reached down here, and there was only a light haze of smoke, which made it easy to search for our missing person. Our hand lights danced around the basement storeroom but revealed no victim.

> *Our hand lights danced around the basement storeroom but revealed no victim.*

Mounting Suspicion

A steel door at the rear led outside, and it had been forced open from the hinge side. Pushing the door to the side, I exited to see a concrete stairway with five steps leading into a little concrete courtyard. The courtyard was well lit by a large security spotlight. At the top of the stairs was the rear door to the restaurant that Bobby had forced open. My first thought was that either he or a firefighter from the second truck company must have forced the basement door.

My mind was whirling, trying to figure this out. Where did the arsonist go? Was the newspaper driver wrong, or did the arsonist somehow escape? I went down and reexamined the basement door. Bingo! There it was, plain as day. The pry marks were made from the inside of the door. Nobody had forced this door to gain entry, because the outside of the door had no marks or scratches. Someone had forced the hinge from the inside when all he had

> *Bingo! There it was, plain as day.*

to do to open the door was to turn two locks. Why would someone do this? The answer to this riddle was easy: The arsonist had entered with keys but wanted to give the illusion that someone had to force entry. I could see that he used a small pry tool, probably a crowbar, and all of the crimp marks were on the inside.

Why did he do it on the inside? I quickly surmised that he was afraid someone would see him in the well-lit courtyard, and he thought it would fool us. This appeared to be an inside job, probably arson for insurance money.

The Chief had already called for an arson investigator. I told him about the smell of gasoline and how hard the fire was to extinguish. Then I walked the investigator downstairs and showed him the rear door. I had already polled the firefighters on the scene, and nobody had touched that door. The investigator took several pictures of the door and seemed to agree with my theory. The Chief remained on the scene after the rest of us returned to quarters.

Now it was close to 5:00 a.m. I changed out my air bottle and sat down to write my reports, being sure to detail all of the information I could in case it would end up in court.

The Trial

Sure enough, about three years later I received a subpoena for an arson trial. Even though I had been at a couple hundred fires since that restaurant fire, details of the incident were still in my head because it was so unusual. When I met with the State's Attorney, I found out that the owner was on trial, charged with aggravated arson and arson for profit.

The star witness was a restaurant busboy at the time of the fire. This was the young man the newspaper driver reported seeing inside the building. The busboy was badly burned that night and spent several months in a burn unit. He testified that his boss, the owner, had hired him to assist in torching the failing business. The owner had his employee pour gasoline around the restaurant while the owner forced open the basement door to make it look like someone else had done it. He intended for the two of them to light the fire as they were leaving through the basement door, but he hadn't counted on a pilot light from the fryer igniting the volatile gas fumes.

The boy somehow made it down the basement, where his boss beat out the flames. The owner drove him to a nearby hospital and let him out at the door, ordering the boy to tell the hospital that his car battery blew up in his face.

Next it was my turn to testify. As I took the stand, a 6-foot-high movie screen behind me was showing my fire reports, blown up for all to read. Thank goodness I had taken the time to write a neat, accurate report. I make it a practice of preparing an especially detailed report of any incident involving damage to property, injuries or deaths, and suspicious starts.

After the State's Attorney took my testimony, the restaurant owner's high-priced attorney attempted to create doubt in the jury's mind. He wasn't successful with me because, fortunately, I was a student of the job. I had taken several forcible-entry classes, had years of experience, and I had taught several classes on this subject. Also, the pictures and my report backed up my statements. The defense attorney quickly dismissed me.

After two more days of trial, the State's Attorney called me with the verdict. On the last day the owner had brought in his six children, who all sat in the front row and cried in front of the jury. After a lengthy deliberation, the jury came back with its finding: not guilty.

I couldn't believe it! The criminal justice system doesn't work all of the time. I could only take solace in the knowledge that we did our part. That's all we can do.

Lessons Learned

- **We always should look for the cause of the fire and preserve any evidence of arson.** Furthermore, we should mentally note any evidence that we must disturb while extinguishing the fire.

- **We must stay at the scene so we won't break the chain of evidence.** We can't leave the scene of suspected arson until police or fire investigators arrive.

- **All findings must be documented.** As in this story, we may end up in court with our reports. That's why we must be sure that they are neat and all findings are well documented.

- **Arson fires = danger to firefighters.** Any suspicious signs such as the smell of accelerants, unusual smoke, or stubborn fire should heighten our safety concern. The arsonist may have weakened the structure or used a lot of accelerant. Your primary concern must be firefighter safety. Utilize risk management.

Discussion Questions

1. In addition to the clues in this story, what signs of arson might firefighters find at fire scenes?

2. What should we do to protect evidence arson once we find it?

3. Why does arson usually make the building fire more dangerous to firefighters?

Fitting In

Lieutenant Julius Stanley
27 Years in the Fire Service

I sat there, a senior firefighter at the age of fifty-one reflecting back to when I was a twenty-three-year-old graduate of the Fire Academy. My first assignment was to the jet boats during the boating season. These were high-speed boats that patrolled the shoreline for the safety of boaters, swimmers, and beach-goers during the warm months. In the fall and winter months I was detailed as a relief fire-fighter to various firehouses in the first district.

Learning on the Job

About a year into my career, I was detailed to Truck 5, which was housed in Engine 18's busy firehouse. This house opened in 1872, and was one of the oldest firehouses in the city. This was my first detail to Truck 5, and I enjoyed the historic feel of this firehouse and its location, which ensured plenty of work. This assignment was entirely different from my experience on the boat, and I had never been to a big fire.

Adding to my discomfort, in my relief role I worked at different houses and with new people every day. Each day I started my shift feeling like I didn't quite fit in. Also, in 1979 the Department was still adjusting to the increased numbers of African Americans, myself included, coming on as the firefighters.

One thing I noticed was that even though the rule book said that our twenty-four-hour shift would start and finish at 0800 hours, the firefighters expected to be relieved by the incoming shift at 0700 hours. I made sure to always arrive before 7 o'clock.

It became obvious, too, that hard work and effort was what earned the respect of the firefighters in this close-knit community. I worked hard to demonstrate my willingness to work hard. When everyone joined in on the scheduled house and apparatus work, I assisted in whatever had to be done, and always pushed to do more on my own. I stayed quiet but I always had my eyes and ears open. After all, I was still a rookie learning to be a firefighter.

> *I worked hard to demonstrate my willingness to work hard.*

Opportunity Calling

My learning experience was about to take on a life of its own. At the sound of a piercing bell in an otherwise eerie silence, everyone rushed to the rigs. Firefighters slid down the pole, ran toward the rigs, kicked off their shoes, and jumped into their boots. Apprehensively, I followed, not knowing what to expect. Even though I had been well trained at the fire academy, those days seemed far removed as our truck sped through the streets.

Then came the words I had been waiting to hear: "We got a hit!" (a working fire). My eyes and my sense of smell confirmed those words. Facing backward in the jumpseat, I smelled it before I could see it. We entered a street that was cloaked with smoke—so much smoke that I couldn't see where the street ended and the sidewalk began.

As soon as we stopped, I jumped off the rig and grabbed my pike pole and axe. I had to run to catch up to the lieutenant, who was disappearing into the whitish-gray smoke. He wasn't the regular assigned officer. Like myself, he was detailed here for the day. At morning roll call he had instructed me to stay with him, and now I could see that this might be easier said than done. I had to hustle to catch up with him at the door of a large factory building.

He placed the adz of his halligan bar just below the top hinge and instructed me to give him a bight. Using my axe, I struck the halligan bar several times, burying the adz deeper into the jamb until he told me to stop. He pried the three hinges away from their seats, each in the same manner. Soon the large wooden door was free from the jamb and heavy gray smoke was rolling out at us. We knelt to the side, let the smoke lift a little, then slid inside so we could locate the seat of the fire.

The first breath of smoke told me to stop: Don't go any farther. I tried to slow my breathing, and I had been taught that the best air would be found at the floor. When I squatted and took short breaths, I was relieved to find that I wasn't suffocating. The Lieutenant's light was in front of me, and I stayed on his tail as he had instructed me to do. I could hear striking sounds to the left of the light and realized that he was tapping the wall with the halligan bar.

For the first time, I realized we were following a wall, doing a left-hand search. Occasionally we left the wall to go around an object, but we returned to the wall immediately. The light in front of me stopped. I felt heat on the side of my face and crawled up behind the kneeling figure in front of me. The roar was deafening, with loud popping and crackling sounds and the distinct sound of rushing air. I moved around the lieutenant to take a look and felt like I was starting to melt. He pulled me back behind the safety of the wall, and I noted to myself: "Next time stay behind the light!"

> *The roar was deafening, with loud popping and crackling sounds and the distinct sound of rushing air.*

"Follow the wall back to the door, and lead the engine company back here," he instructed me.

With my hand against the wall, I started back in the direction we had come. About halfway to the door a light was heading in my direction, and I could hear voices and coughing. It was the engine company leading out a hose in our direction. How did these guys know which way we had gone?

"Over here!" I waved my light, directing them toward the fire. Three firefighters dragging a hoseline passed by me, heading toward my officer. I followed the last one.

We gathered along the wall, which protected us from the fire, and the engine officer yelled, "Send the water!" The firefighter in front of me ran past, following the hoseline to the door. I helped the other two stretch out the hoseline. While we cradled the line, waiting for the water to surge through, I could hear striking sounds high above. Then glass from skylights crashed to the floor around us, and we could hear the scream of a K-12 saw in the distance. Then came the buckling sound of the hose being charged with water. The air briefly turned misty as if in a steam

> *Glass from skylights crashed to the floor around us.*

room. Other lines were deployed at this incident, but ours was the only line at the seat of the fire.

When the fire was out and the interior was vented, the visibility inside the building slowly returned. We picked up the hose, surrounded by large puddles resembling a swamp on the concrete floor. I had been taught that when we can't see the floor, we should push a tool out in front of us like a blind person's cane. I stabbed at the concrete in front of me with my pike pole as I walked through the puddles. On one stab I felt the pole sink and stopped in my tracks to investigate. I pushed my pole down as deep as it would go. At about 6 feet, I couldn't feel the bottom.

"Watch out! There's is a big hole in the floor," I yelled to the lieutenant. Slowly I felt out the perimeter of the pit. It was almost 10 feet square, and more than 6 feet deep. He kept me in place to warn others while he went and informed the Chief. The puddle of water was concealing a large tank that was used to process metal. It could have been a death trap to a firefighter wearing hip boots and heavy equipment. After the last firefighter exited the building, I left my post and rejoined the others outside.

The Reward

I was helping the others carry the hose to the engine when the lieutenan called me over to the truck. "Put it back in its bed," he said, motioning toward the aerial ladder.

Everyone stopped in their tracks, knowing that this was an honor reserved for firefighters assigned to that truck. You could've cut the tension with a knife. I climbed up onto the turntable and replaced the assigned firefighter at the controls. I couldn't figure out the controls at first, so the driver instructed me, and I put the ladder to bed. I felt a pride and a sense that I belonged to something much greater than its parts. At the same time, I silently thanked the

> *I felt a pride and a sense that I belonged to something much greater than its parts.*

lieutenant, and acknowledged the disappointment of the firefighter who had to swallow his own pride and quietly allow me to bed his ladder.

A Student of the Game

On future details to Engine 18's house, I was treated like a family member. I understood that race wasn't a factor in my acceptance. It was my willingness to learn. I had been sent to the boat directly from the Fire Academy and developed skills in scuba diving and steering a 42-foot twin diesel jet boat, but I had little experience in fighting fires.

The lieutenant had to take special care to ensure my safety until I learned the ropes of this dangerous game. The knowledge, skills, and aptitude for the job were not going to come from just books alone. I would have to pay my dues on the street. I became a student of the game and learned each day from every incident. I recognized the value of combined experience and education as the core of what it takes to be a firefighter.

The men and women of this Department helped mold me into what I am. I experienced hurt, life-changing shame, growth, pride, faith, understanding, and true brotherhood. To borrow a line from the Executive Development Institute, "All that I am, I owe. I live eternally in debt."

Lessons Learned

- 🏵 **The respect and trust of our fellow firefighters have to be earned.** Hard work and determination form the path to respect.

- 🏵 **It helps to be humble.** The best way to demonstrate humility is to enter the fire service willing to learn and to help our fellow firefighters.

- 🏵 **Safety is every firefighter's responsibility.** We have to be on the alert at all times for hazards, and to warn everyone on the fireground when we encounter them.

Discussion Questions

1. What advice would you give a brand new firefighter at the beginning of his or her career?

2. What is the best way for a new firefighter to gain the acceptance of his or her company?

3. What avenues are available in your area to enhance your education in the fire service?

Earning Respect

Engineer Judy Brosnan
18 Years in the Fire Service

Rookie firefighters fresh out of drill school often feel the pressure of having to prove themselves to other firefighters. As a female firefighter entering the Fire Department in 1986, I felt this pressure—and more. Prior to my class of recruits, the Department had only one female firefighter. Our graduating class had twenty women in a class of 150. I think all twenty of us felt like we were under a magnifying glass.

This pressure was short-lived. Most of the firefighters readily accepted me into the Department family. One of the other recruits, Frank, and I were assigned to Truck 13, a truck company with outstanding firefighters and a good officer. He was tough but fair, and he was a good teacher, too.

The First Year

The first year was a tremendous learning experience and flew by. Each day that we trained and drilled together, we bonded more as a team. To shape us into well-rounded firefighters, our officer, lieutenant Bennett taught us every aspect of truck work. Our knowledge base increased day by day, fire by fire.

In the beginning, Frank and I were under the direct supervision of Lieutenant Bennett. We were at his side doing forcible entry, search and rescue, ventilation, overhauling, and salvage work. Soon he entrusted us to work in teams with more experienced firefighters. After about a year, he occasionally sent me to the roof for vertical ventilation with one of the other firefighters.

Testing My Worth

One afternoon a call came in, and from a few blocks away, we could see a thick black cloud of smoke rising into the gray winter sky. "We got one!" (working fire), our lieutenant yelled. I was excited because at morning roll call I was assigned to roof ventilation along with Rick, our driver.

While Rick spotted the truck in the front of the building, I studied the conditions. It was a huge, three-story courtyard apartment building with fire showing in at least one apartment on the second floor. Flames were blowing out of two windows near the center courtyard, and thick smoke filled the sky. We had to ventilate over the stairway and third-floor apartment right away. Smoke and superheated gas were filling the structure, and with at least thirty-six units, there was the potential for a lot of trapped occupants.

Rick was an experienced firefighter with six years on the job. As he swung the 100-foot aerial ladder onto the roof, I grabbed the K-12 circular saw, an axe, and an 8-foot pike pole. He quickly had the ladder in position, swung the saw over his shoulder, and started up to the roof. I followed closely, carrying my axe and pike pole.

"This is what it's all about."

On the way to the roof, I saw the engine company leading out its hose into the center doorway, and lots of scared occupants running out the door. "This is what it's all about," I thought.

When Rick and I reached the flat asphalt roof deck, we had to walk about 100 feet to reach the fire area. I followed him, now carrying his axe as well as mine so he could ready the saw. We walked along the outside wall, knowing that outside walls are the strongest parts of the roof deck. As soon as I stepped over a 1-foot firewall onto the section of roof over the fire, something didn't feel right. The roof felt spongy, and this was the first time I had ever felt anything like this. In training we had been warned that roofs with a spongy feel were close to failing.

A Close Call

Suddenly—crash!—the roof deck collapsed and Rick disappeared into a hole. As he was falling into the fire, he reacted quickly, throwing the

50-pound saw over a firewall onto the next section. He threw one arm over the coping stone on top of the wall and was dangling over the fire.

Instinctively, I dropped my tools and dived onto the wall. I crawled to where Rick was and grabbed hold of his arm: "Hang on—I've got you." He was a big, muscular guy, well over 200 pounds, so it was a struggle to hold him.

Suddenly, another firefighter grabbed me by my thigh. Another truck company had just reached the roof and saw our predicament. Rick was able to throw his other arm over the wall and finally help pull himself back onto the roof.

Within seconds, flames were shooting out of the hole, and our job on the roof was done. The fire had vented itself, and now it was time to regroup with our company inside.

Our teamwork at this incident had been successful. All of the occupants had been safely evacuated, and the fire was extinguished without any firefighter injuries.

After the fire, we talked about our close call, and I realized I should have trusted my instincts and spoken up about the spongy roof. I had assumed that because Rick was more experienced, he would have known about the spongy roof. In fact, he hadn't noticed it. Thankfully, the firefighter from the other company had grabbed my leg, keeping me from falling in and also providing more support to Rick as I clung to him.

Postscript

About a week or so later, I was at a firefighter's retirement party and a guy tapped me on my shoulder, offering his opinion of women in the Department: "You girls don't belong on my job." Just seconds later, someone tapped me on the other shoulder. It was the firefighter who had grabbed my leg on the roof: "Hey, you did a great job last week, and I'm proud to work in the same Department as you!"

That incident made me realize that in life there are "people who row the boat" and "people who rock the boat." The ones who matter are those who row, and most of my fellow firefighters row the boat.

Lessons Learned

⚖ **We must heed the warning sign that spongy roofs are ready to fail.** The spongy feel of the roof deck in this story was caused by intense fire in the cockloft area. The fire compromised the 2-by-10 joists of the ordinary-construction roof deck. The weaker the joists get, the spongier the roof will feel. If we notice any change in the feel of the roof, we must immediately evacuate.

⚖ **The strongest part of the roof is near the sidewalls.** In ordinary construction the roof joists tie into the masonry walls at the narrowest dimension, which usually is a sidewall. As we cross a roof deck, we should walk along sidewalls whenever possible.

⚖ **We should trust our instincts and warn others of dangers we see or feel.** Sometimes more experienced firefighters might miss something that we see. When we detect a dangerous situation, we should immediately relay this information via the radio to the Incident Commander so all firefighters on the scene will be aware of the situation.

Discussion Questions

1. What is one warning sign of roof collapse?

2. Other than collapse, what are some hazards of vertical ventilation?

3. What are some ways to fit in and get accepted when joining a fire department?

The Day That Changed My Life Forever

Although twenty years have passed since this tragic incident, I still remember every detail, every sound, every smell, every conversation of that day. I remember the weather, the food we ate, the jokes we told, but mostly I remember the terrible sights I witnessed. These visions are ingrained in my memory, and I have relived that terrible night again and again in my mind. It is a nightmare that I can't escape.

Looking back at my life, my memories are in living color, except for one. Memories of that day are in black and white, as if shrouded in a dark veil. One day, January 31, 1985, changed my life forever.

Retired F.F. William Karda
21 Years in the Fire Service

A Routine Morning

The day began like any other. I awoke at the usual time, followed my regular morning routine, and left for work at the same time I always did. Light snow was falling as I drove through the predawn darkness on my way to work. The new snow added to the four inches already on the ground, which made the roads slick. "It's a good thing the traffic is still light at this early hour," I thought.

I walked into the firehouse at my usual time, about 6:10 a.m. For about four years I had been assigned to Truck 58, housed with Engine 7 and Ambulance 7. Every day started the same way. As usual, I was the first to arrive that morning. The first thing I did was to load my gear onto the truck and check out my SCBA (Self Contained Breathing Apparatus). Then I sat back in the kitchen, poured a cup of coffee, sat down to read the paper, and waited for my fellow firefighters to arrive.

My captain, as always, was the next to walk through the door. He smiled and said, "Hey, Billy, what do you do, live here?" After putting his gear on the rig, he joined me at the table. He wasn't just my boss. Cap was my friend, and I always enjoyed our conversations.

Shortly before 7:00, Mike walked in, carrying a gym bag.

Cap commented, "It looks like you're having a bad morning."

Mike replied, "I am. There should be a law against working on birthdays."

After the revelation that it was Mike's twenty-sixth birthday, Tal, Sam, and the guys from the engine walked in and the kitchen came alive. The large kitchen table was soon surrounded by firefighters. With this group there was no shortage of laughs.

It was my day to be the cook, and I sat at the table collecting money from the guys and making a shopping list, including a cake for Mike's birthday. Around 8:30 the group started the daily house and apparatus cleaning, and I left for the grocery store. By 10:30, I was preparing lunch. On this day lunch would have to wait.

Lunch on Hold

Around 11:00 the speaker crackled: "Engine 7, Truck 58, take in the still alarm." (At that time it consisted of two engines, one truck company, one rescue squad, and one battalion chief.) I quickly turned down the burners. The sliced beef would have to simmer longer than normal. Both rigs rolled out the door within seconds, lights flashing and sirens wailing.

The first to arrive at the scene, we found a large three-story apartment building with no signs of fire. Mike spotted the truck to line it up with the center of the building. From this location we could ladder any portion of the building if needed.

There was no smell of smoke, so the captain split us up into groups to investigate the apartments to the west. Engine 7 headed toward the apartments on the east. Cap and I investigated the basement, which was clear—no signs of fire. This appeared to be a false alarm. Just then the captain's radio crackled. Engine 7 had found a fire on the opposite side of the building, up on the second floor.

We came out of the basement, gathered the rest of the company, and ran around the whole building to reach the fire area. By now, the Chief, Squad 2, and Engine 91 had arrived. Engine 7 was leading out a hoseline to the rear, and two firefighters from Squad 2 were raising our main ladder to the roof. The Chief wasn't too happy about this and questioned our captain; "Why is the squad raising your ladder for you?"

The captain tried to explain that nothing was showing when we pulled up, but the Chief wasn't interested. "Take a two and a half up the front stairs to the second floor," he barked.

Cap didn't look too happy. He ordered Mike to the roof to help the two squad guys. The rest of us took a 2½-inch line off Engine 7 and led it up the front stairs.

We stretched the line up to the third floor landing and looped the hose back down the stairs to the door of the fire apartment. Heavy, dark-gray smoke was pumping out from the cracks around the door. Cap gave the order over the radio to send the water. As the hose sprang to life, I cracked open the pipe (nozzle), and it hissed as it bled off the air. Within seconds water flowed in spurts for an instant, then in a solid stream. I closed the pipe and we were ready to attack the fire.

My partner, Sam, was standing next to the door. "I'm ready," I said, and Sam forced open the door. As the door flew open, heavy smoke and flames rushed out to greet us.

Game Time

"It's game time," I told myself as I started to crawl in with the hose under my arm. Sam was second up, and I felt his shoulder push into my back as we moved the big line forward. I could hear the Engine 7 firefighters working their line, coming toward us from the opposite direction. It sounded like a war zone as our fire streams battered their way through the apartment. Finally, room by room, the fire was extinguished.

It sounded like a war zone as our fire streams battered their way through the apartment.

All of the firefighters in the fire apartment performed a primary search while the fire was being brought under control. The smoke was starting to lift as we did a secondary search. We now had too many firefighters in the apartment and were getting in each other's way. Engine 7 was still hitting hot spots with their hoseline, Truck 35 was pulling ceiling and sidewalls, and we were trying to do a thorough secondary search. We had almost completed our secondary search when Sam stepped on a debris pile behind the front door and felt something unusual. "I got one," he yelled.

We quickly dug out an adult male from behind the door. After getting him out of the apartment, we started CPR until an ambulance company took over. Later we found out that he didn't make it.

While we were picking up our hose and tools, the Chief questioned our captain as to why we had taken so long to find the victim. We could tell that Cap was angry, and we hoped that it wasn't at us.

Post-Fire

Back at our quarters, while we were eating lunch, we critiqued the fire. What could we have done differently? The captain assured us that we did everything we could under the circumstances. He told us he was proud of us, and that meant a lot.

> *He told us he was proud of us, and that meant a lot.*

We had only one more run before dinner, and it turned out to be nothing. We celebrated with a dinner topped off with a birthday cake for Mike. Afterward, the whole house sat around the table talking and throwing digs at each other. Cap told us that this was his twenty-fifth anniversary with the Fire Department. We congratulated him, then blasted him with a few "old man" jokes. Had I known it was his anniversary, I would have bought him a cake, too.

Mike was scheduled for first watch at 10:00. Because it was his birthday and I wasn't tired, I stayed up and took it for him. It looked like it was going to be a quiet night. At about 1:30 I woke up one of the engine guys for watch, and I went to bed.

A Big One

Around 3:30 a.m. the silence was broken, bells rang, the bunkroom lights came on, and ten firefighters sprang to life and ran for their rigs. Within seconds we were heading to a still and box alarm (second alarm) on Milwaukee Avenue. We had traveled only a short distance when we heard an excited Battalion Chief calling for a 2-11 alarm (third alarm).

We knew a big one was waiting for us. From a few blocks away, we could see the flames lighting up the night sky.

We knew a big one was waiting for us.

While we were approaching the fire scene, the Chief called us on the radio: "Truck 58, position your rig on the west side of the fire building." Without hesitation, we pulled into a parking lot to the west of the building, just out of the potential collapse zone. Heavy smoke and fire were pouring out of an electronics store that was about 40 feet wide by 125 feet deep. Two floors of occupied apartments were above the front 50 feet of the store. The rear 75 feet was one story. Heavy smoke was cascading out of the apartment windows on the second and third floors.

As we climbed out of the truck and grabbed our tools, the Chief came over to us. Engine 7 was on the radio requesting a truck company in the rear to force entry. Cap said, "We'll get that, Chief."

"No, take your company up and open this roof." The Chief pointed at the one-story roof over the rear of the store.

"Throw a ground ladder over there," Cap ordered us. We quickly grabbed a 28-foot ladder and raised it in the snow-filled parking lot. I held the ladder as, one by one, our company climbed to the roof. When the captain stepped off onto the roof, he held the top for me. I started up, axe and pike pole in hand.

When I reached the halfway point, the Chief called me back: "Karda, come here for a minute."

I shouted to the captain that I would join them when I finished whatever the Chief needed.

When I came down, the Chief ordered me to go up a 20-foot, straight-frame ladder that was already in position and open two second-floor windows that were covered with plywood. I left my axe at the base of the ladder and quickly ascended to the second-floor windows, pike pole in hand. Hooking the ladder with my leg, I used both hands to pop the plywood in with my pole. I expected heavy smoke to push out of the openings, but very little smoke exited. When I looked inside, heavy black smoke was churning. "This doesn't look right," I thought.

Back on the ground, I told the Chief what I had seen. He seemed preoccupied and told me to go up and rejoin my company.

> *Then came the blood-curdling screams that haunt me to this day.*

Grabbing my axe and pike pole, I started back up the 28-foot ladder. Then I dropped my tools onto the snow-covered roof, grabbed the tips of the ladder with both hands, and swung my feet onto the roof deck. Before I could even turn to face the roof, I felt movement in the deck. Then came the blood-curdling screams that haunt me to this day.

The Roof Collapse

Turning, I saw my whole company sinking into the roof about 30 feet in front from me. Helpless, I watched in horror as the whole center of the roof collapsed and swallowed up my brothers. As they disappeared into the hole, flames shot up around them. I felt the roof shift below my feet and, dropping my tools, I dived back toward the wall. I landed on top of the wall and grabbed onto the slippery, clay tile coping. For a split second I clung there, then jumped over the roof edge and landed in the parking lot about 20 feet below. My fall was broken as I landed feet first in a 3-foot bank of snow left by snowplows.

When I heard screams, I looked up to see two firefighters from Truck 14 clinging to the wall about 20 feet away from the ladder. Their axes and pole were dangling over the wall. I yelled, "Drop your tools and hold on," I shouted. Then, with the help of Tom, a firefighter from Squad 2, we started rolling the ladder in their direction as they inched their way along the top of the wall toward us. Screams came from all directions.

I held the ladder, and the firefighters climbed down. When the first firefighter reached the base of the ladder, I grabbed him and asked, "Did you see any of my guys?"

He looked me in the eye and said, "I think they're all gone."

I ran for the truck and grabbed a battering ram out of a compartment. Others joined me and we started ramming the tool into the solid brick wall. I knew where the guys were, and I thought we could get to them. Unknown to me at that time, others had seen them fall through the roof, and rescues were being attempted from several areas. Flames now were shooting 40 feet into the air through the collapse area.

After several minutes of working the battering ram, an officer came and got me, saying the command van needed me. In a daze, I followed him toward the van, where

> *Flames now were shooting 40 feet into the air through the collapse area.*

some Chiefs questioned me as to what I had seen and what my company's location was when the collapse occurred.

I could tell that they didn't believe anyone survived because they were talking about recovery, not rescue. Everything was starting to hit me. One of the Chiefs ordered me to go and get checked out by paramedics. Someone walked me over to an ambulance.

Who Survived?

In the ambulance, one of the medics told me that Sam had made it out. He had fallen through the roof with the other three guys and rode the roof down into the fire. When he hit the ground, he saw a roof timber angling up toward the roof. Completely engulfed in flames, he grabbed this timber and climbed out of the inferno. Still on fire, he was grabbed by the lieutenant of Squad 2, who rolled him in the snow to extinguish the flames. All other rescues were hampered by secondary collapses and intense fire.

When I was told that Sam was in the ambulance next to the one I was in, I demanded to see him. The medics tried to stop me from leaving my ambulance, but finally one of the medics walked me over there. If Sam had survived, I thought, maybe others had, too! When I opened the side door, three medics were working on him and wouldn't let me in. I saw him lying there, and my heart sank. He was so badly burned that I couldn't even recognize him. The medics took me back to the other ambulance.

During my exam, I heard my sister's voice screaming for me. A fire-fighter led her to me so she could see I was safe. She lived only a half-block away, in the same house where I grew up. She and other family members had come down the block to watch the fire. When the collapse happened, they heard that firefighters were missing and became alarmed. I sat up and talked to her, assured her that I was all right, and I asked her to call my wife Pat and let her know that I was okay. This was good timing, because at home my wife had just awakened to news on the radio about the collapse and the report of missing firefighters. She had just called her dad for a ride to the scene when my sister called to say I was all right.

Everything was starting to sink in.

It was now about 6:00 a.m. The exam showed that physically I was fine. Emotionally, though, everything was starting to sink in. One of the Deputy Chiefs on the scene told the ambulance to take me back to the firehouse. I had never felt so confused as I sat, alone and dazed, in the dimly lit firehouse kitchen. One by one, the second shift members started to come in, full of questions; it was painful. They had to grab their gear and report to the fire scene. On the way to the building, one of the guys drove me home.

Home at Last

It was an emotional homecoming with my wife. My fourth-grade son, Ed, had gone to school against his wishes. He knew there had been a bad accident, but my wife assured him that has dad was okay and she sent him off to school.

The phone and doorbell started ringing. The press had found me. My wife screened them out, knowing that I wasn't up to answering their questions. A Deputy Chief came by to again question me on my company's location. Only two bodies had been recovered, and the Chief took me to the morgue to identify them. It was the hardest thing I have ever had to do in my life—to look at the charred remains of my brothers. I could identify one of them only by a piece of jewelry. I couldn't tell who the second one was.

When we got back to my house, I could hardly wait to see Ed. It was still too early to pick him up, but I couldn't wait any longer and walked with my wife to the school. Other parents wanted to ask me questions, but I couldn't talk about it, and my wife kept them at bay. When I looked in the window of my son's room, he saw me, jumped up, and ran into the hall screaming, "Dad! Dad!"

We hugged each other hard. He looked up at me and said, "Dad, I thought they were lying to me. I didn't believe you were all right." I couldn't wait for the final bell, so I took him home early.

Back to Work

The rest of that day and the next I was in a fog. Then Sunday morning arrived and at 5:00 a.m. my alarm clock went off. It was time to go to work. I kissed my wife, hugged my son, and out the door I went. Once again I drove through the predawn darkness to get to work. I walked in the back door at 6:10 like I always do. I loaded my gear on the truck and checked my SCBA, and then, like always, went into the kitchen. I poured a cup of coffee, sat at the table, and waited for my buddies to come in. Cap should be coming through the door any minute, followed by Mike, Sam, and Tal. Of course, they never did.

Sam would never make it back on the job. He spent months in the Burn Unit. His stay there was long and painful, because he required many surgeries and months of rehabilitation. Finally, some six months later, he was ready to go home. We made arrangements to take our company out of service for a few hours and take Sam home the way he deserved, on board Truck 58. At the hospital he received a tearful, yet happy goodbye from the doctors and nurses, who had formed a permanent bond with Sam. We loaded Sam's bags on the truck, helped him into the front seat, and I proudly drove him home.

When we rounded the corner onto Sam's block, I turned on the lights and siren to honor our hero. It was a sight I'll never forget. Well over a hundred neighbors and family members lined the streets, holding "welcome home," "we love you, Sam," and "we're proud of you, Sam" signs. He was smiling from ear to ear, and there wasn't a dry eye on the block. It was a fitting homecoming.

It was a sight I'll never forget.

After years of surgeries, Sam bought and managed a beauty shop. In his "after-fire" life, he has led the happy life he deserves.

Dealing with losing my whole company has been a life-long struggle. After twenty years, the only cure I've found is time. I left the job ten years later with health problems, but I have recovered fully. The support of my family, especially my wife Pat, helped me through those terrible times.

This mural was painted on the wall of a neighboring bank building to commemorate the three fallen firefighters. The mural and a small park dedicated to the firefighters mark the spot where these heroes gave their lives. Photo courtesy of Dave Berger.

The support of my brothers and sisters in the Department was a big help, too. The fire service is truly a big family.

After the incident, I spent almost a year making a memorial to honor these three brave heroes. It was therapy for me, and a way to ensure that these three great men will never be forgotten. This memorial hangs in Truck 58's quarters. They will live forever in our hearts and our minds. God bless.

Lessons Learned

● **Critical incident stress debriefings are crucial.** Since the incident in this story, twenty years ago, much has changed. At that time, we would witness tragedies and then just go home when our shift was over and continue on with our lives. Most of us never even talked at home about the terrible things we witnessed. We held everything inside. Today we deal with the mental trauma caused by these experiences.

● **It is our job to look out after each other.** Fortunately, Billy Karda received tremendous support from his wife and family. Today, firefighters dealing with such tragedies are offered Critical Incident Stress Debriefings (CISD) and follow-up counseling if needed.

● **We must understand building collapse.** The biggest factors leading to collapse are the type of construction and the time of fire involvement. This building was ordinary construction, (masonry walls and wood joist construction), but it had a deeply seated fire. The longer a building burns, the weaker it gets. A collapse usually is triggered by impact load, which could be a firefighter walking or working.

● **We have to watch out for extra weight on roof decks.** The members of Truck 58 were next to a large air-conditioning unit, which added hundreds of pounds to the roof deck. This roof also carried the weight of four inches of snow. This added weight, along with the weight of four firefighters, sped up the roof's failure.

● **We must be especially alert to arson fires.** This fire proved to be arson for profit. Arson fires are especially dangerous because of the use of accelerants, which spread the fire dangerously fast. Also, arsonists have been known to weaken structures to speed the collapse.

● **Melted snow indicates the area directly above a large volume of fire and signals us to use caution in these areas.**

Discussion Questions

1. While reading a building during size-up, what are some indicators of collapse?

2. What warning signs usually precede the collapse of a roof deck?

3. Does your fire department offer Critical Incident Stress Debriefings (CISD)? How important is this service?

Not This Time!

Battalion Chief Steve
Chikerotis
27 Years in the Fire Service

It was April of 1985; three months had passed since the tragedy on Milwaukee Avenue that killed three brother firefighters. We tried not to think about it, but it was always in the back of our minds, reinforcing the reality we face daily. Every morning when we leave for work, we know it could be the last time we will see our families. All we can do is to make sure we hug them, tell them we love them, and then go off and do our job.

Death and serious injuries to our comrades don't make us less aggressive, but they do make us more safety-conscious. At the time, I was a firefighter on Rescue Squad 2. The day after the tragedy on Milwaukee Avenue we moved into our new firehouse, Engine 91's quarters on Pulaski Avenue. It was a big new house with a superb group of firefighters. As soon as we moved in, I felt at home.

This group was one of the best ever assembled. We took our jobs seriously, but not so seriously that we forgot how to laugh. If the guys weren't laughing with you, they were laughing at you. This was no place for the thin-skinned, but it did serve a purpose. The laughter kept our minds off the ugly reality we face daily. As

We took our jobs seriously, but not so seriously that we forgot how to laugh.

one of the busiest fire companies in the city, when we came to work, we expected to go to fires. We worked hard, and we also trained hard. We learned a lot from our captain, Big Bill, and we learned a lot from each other.

The Team in Action

It was about 3:00 p.m. on a sunny day, and so far the day had been slow. We were sitting around drinking coffee when the voice on the speaker announced, "Squad 2 and 272 take in the still alarm" (first alarm). Coffee mugs slammed to the table and everyone ran for the rig. Within seconds we were out the door and heading north. The address was less than one mile away. Engine 7 and Truck 58 were already on the scene reporting a fire, and we would be there in a minute.

Looking to the north, a dark black cloud of smoke scarred the powder-blue sky. All eyes were reading the fire building. It was a corner building of ordinary construction (masonry and wood joist), about 40 feet wide and 125 feet deep. The building ran from Pulaski Avenue all the way to an alley in the rear. It was three stories high in the front with two floors of apartments over a liquor store, and one story in the rear. The apartments were occupied, but the liquor store had gone out of business a while back and was boarded up tightly with plywood. Heavy black smoke was pouring out of every crack of the plywood.

John, our driver, spotted the rig on the sector two (left) side of the building. As we got out, our captain ordered, "Steve and Danny, do a primary search of the second-floor apartments."

Reading the building, I told him, "We'll take the rear stairs, Cap."

We headed for a set of stairs to the roof of the liquor store. A path across the roof led to the rear doorway to the apartment building. As soon as we took a few steps on the roof, we could feel that something wasn't right. Our feet seemed to sink slightly into the roof, similar to stepping on a thin sponge. We recognized that a spongy roof is a dangerous sign, usually caused by a tremendous amount of heat below the roof deck, which can lead to a collapse.

Eerie Similarities

To our left we heard a K-12 saw, and through the smoke we could see the crew of Truck 58 in the middle of the roof, ventilating. It struck me as eerily similar to Milwaukee Avenue, the tragic fire from which we were still reeling. An officer and two firefighters were busy working and seemed unaware of the dangerous roof condition. Truck 58 was the same company that had just lost three members.

"Freddie," I yelled to the officer, "this roof is bad. Get them off!"

Through the smoke I saw them scurry toward the rear stairs. Danny and I made our way quickly toward the rear door to the apartments. We chose a path close to the sidewall. This is the strongest part of a roof deck, where the joists tie into the brick wall. When we reached the rear wall of the apartments, we had to cut to the left across the roof to get to the rear door of the apartments. As I opened the door and stepped in, Danny was just a step behind me. My second foot hit the floor, and I turned quickly to look for him. I don't know why, because I don't remember hearing a sound. He was sinking quickly through the roof, as if he had a rug pulled out beneath him.

> *He was sinking quickly through the roof, as if he had a rug pulled out beneath him.*

We reached out to grab each other, and I pulled him in. He ended up on top of me inside the doorway, and we gave each other a "that was close" look. Then I asked, "Do you think the truck made it?"

Danny replied, "Yeah—I saw them make the stairs."

We looked out the door, and the whole roof was gone. Replacing what had been the roof was a sea of flames dancing several feet in the air.

This current photo shows Dan Fabrizio and Steve Chikerotis, now both Battalion Chiefs.

A Close Call

We knew the companies below were having forcible-entry problems because the liquor store was boarded up tight. The roof collapsed so fast that we thought nobody had made it inside yet. This was good.

Helping each other up, we shook our heads at our close call and got back to work, kicking in doors and herding the apartment occupants through the smoke-filled hallway to the stairs that would lead them to safety. There was a lot of commotion on the radio—people screaming and stepping on each other's messages.

Suddenly the radio got quiet, and I heard my captain's voice: "Chikerotis, Chikerotis."

It really got my attention because we normally don't use names on the radio. Was my captain in trouble? Did he need help? I answered, "Squad 2, go ahead!"

"Chikerotis, is that you?"

"Yes, this is Chikerotis. Do you need us?"

"Is Danny with you?"

"Yes, Danny is with me. We're doing primary search on two [second floor]. Do you need us?" "No—carry on," he replied, and we continued our search.

Unknown to us at the time, as soon as Truck 58's crew stepped off the roof onto the stairs, the guys heard a sound and looked back to see the roof let go. They didn't think we made it across in time. Seeing nothing but flames, they retreated down the stairs to the alley below. When the roof failed, the whole deck fell in at once. The displaced air caused the boarded-up windows and doors to explode outward. Flames shot out of every opening, similar to a backdraft. The explosive force knocked several firefighters off their feet.

> *The explosive force knocked several firefighters off their feet.*

Our Captain Big Bill, and two firefighters were forcing the rear door when it let loose. He was checking the well-being of his two shaken firefighters when the truck company from the roof came down the stairs. The lieutenant ran up to our captain and told him we were on the roof when it collapsed.

The captain ran up the stairs to look for us. At the top of the stairs, he saw that there was no roof deck, only flames, which now were rising 40 feet into the air. This is where he was standing as he called my name on the radio. He thought he was experiencing the nightmare of every officer— losing firefighters under his command. With all the commotion on the radio due to the collapse, he wasn't getting through at first.

Our Captain Big Bill is a large mountain of a man, with a heart as big as he is. One of our fellow squad men told us later that he feared that the captain was about to jump into the flames to find us, until he reached us on the radio. After the fire, the captain told me that he was so rattled at the thought of losing us that he had a hard time functioning for the rest of the fire.

Counting Heads

Our group saved several people from the apartments that day, but we took our lumps, and several firefighters received minor burns and bruises. We knew it could have been

We took our lumps.

much worse. We must have had three angels looking out for us. Besides, my instincts were sharper that day because the tragedy on Milwaukee Avenue three months prior was still fresh in my mind. On that terrible day, we learned some valuable lessons that stayed with us.

When I went home the next morning and was greeted by my two sons, two and three years old at the time, I hugged them with watery eyes, thinking how their lives would have changed had we not made it. This thought has stayed with me throughout my career. Every firefighter in the country is a member of a family who will suffer if he or she doesn't make it home.

A Side Note

I wrote the above story almost twenty years after the incident occurred. The very next day after writing it, I worked a shift for a friend of mine. Now a Battalion Chief, I was making the rounds and checking on my fire companies when I ran into a firefighter I hadn't seen in years. We talked and laughed about some old times.

After I finished my business in that house and was walking out the front door, my friend called out, "Hey, Steve, I don't think I ever thanked you."

Expecting a joke, I replied, "Sure. Sure. Thank me for what?"

"No—I'm serious. I don't think I ever thanked you for getting us off a roof just before it collapsed."

A chill went through my body. "I have to sit down," I told him.

He joined me and I told him about my book, and about the story I had written only hours before. Until then, I didn't know the identity of the two firefighters who were with the lieutenant. Each of us in the fire service has a chance to touch so many lives. Always remember how important it is to be a student of the job, and always watch out for your brothers and sisters. After this conversation, I knew this book was meant to be.

Lessons Learned

🏵 **Not only truss construction fails; ordinary construction can also collapse.** Firefighters too often lower their guard when they recognize ordinary construction. The fire was burning in the boarded up liquor store for a long time before it was discovered. Buildings don't get stronger when they are burning. Every second of burning brings the building closer to collapse.

🏵 **There is a twenty-minute rule for ordinary construction.** As a general rule, a building of ordinary construction burning for that long is probably close to collapse and we must look for warning signs.

🏵 **Here are some of the warning signs of collapse:**

- *Spongy roofs* serve as a warning sign that there is a large amount of fire in the cockloft area. Do not just evacuate the roof area. Relay the message to the incident commander so others below are warned. *Roof decks are stronger near the sidewalls.* On flat roofs, the location where the joists tie into the walls will be the strongest.

- *A large volume of fire* tells us that the fire has been burning for a while. This is usually the only indicator of how long it has been burning. The larger the volume of fire, the weaker the building.

- *When any portion of a building collapses, loads shift and other structural components become stressed. During collapse all of the firefighters, both inside the building and outside, are endangered.* Large deck collapses can displace the superheated air inside and drive it out windows and doors with explosive force. We have to always expect the unexpected and know what is above and below at all times.

- *We must continuously monitor conditions on the fireground.* Conditions are constantly changing, so our size-up also has to constantly change, too.

🏵 **The instincts we develop and the lessons we learn can guide us.** If we feel that something is dangerous, it probably is.

Discussion Questions

1. Are the instincts we develop throughout our careers important? If so, in what ways?

2. While reading the building during size-up, what conditions triggered the instinct of a probable collapse?

3. What is the importance of communications on the fireground?

"Bring 'em All Home"

Retired Deputy District
Chief William Burns
39 Years in the
Fire Service

From the moment I conceived the idea of this book, one interview I was really looking forward to was with my old captain, Bill Burns. The interview did not turn out fruitful, though, because the Big Guy (our nickname for Bill—he's 6'6" and weighs 280 pounds without an ounce of fat) is not a good interviewee. His modesty prevented him from talking about himself. The stories he told were about other people. Mostly he talked about us, his former firefighters who worked under his command. What he failed to realize is that we are what he made us. I ended up writing the story below in my words. This one is from the heart.

When I came on the job, Big Guy was one of my instructors. At the time, he was a young lieutenant on Engine 77, and he took a three-month detail to the Fire Academy to teach our class. From the first day, he made quite an impression on us recruits. He was like a John Wayne character. A leader, he was a big, strong veteran of the Marine Corps who led by example. He wouldn't ask us to do something that he wouldn't do. He was willing to get his hands dirty to demonstrate the correct way to do something. Big Guy never talked down to us, and he always had time for our lowly group of candidates (new recruits). The traits that impressed me most were his passion for the job and his compassion for the citizens he served.

A Welcome Invitation

After graduating from the Fire Academy, I was assigned to a busy south-side engine company, and Big Guy went back to his westside engine company. About a year later, much to my surprise, I received a phone call from the newly promoted Captain Burns. The Department was forming new rescue squads, and he was to be captain of Squad 2 on the busy westside. He asked, "Would you like to come work for me on Squad 2?"

This was like Christmas, New Year's, and the Fourth of July all rolled into one.

Are you kidding me? This was like Christmas, New Year's, and the Fourth of July all rolled into one. Without hesitation, I responded, "Yes!"

Over the next nine years I worked for Captain Burns, he shaped and molded our young group of firefighters into one of the best companies in the city. Here are some of the lessons he taught us:

- **Enthusiasm.** Captain Burns came to work enthusiastic every day, and enthusiasm is contagious. Our group of firefighters looked forward to coming to work each day.

Captain Burns and his young crew, left to right: José, Steve (the author), Bill Burns, Joey, and John.

- **Pride.** Big Guy took pride in everything he did, and when we did a good job, he praised us publicly. As a result, we worked hard to be the best. We didn't want to let him down.
- **Built-in training.** Captain Burns had a knack for training and drilling us without our even realizing it. We would be having fun going over a new tool or procedure, and before we knew it, we had spent hours in training. He encouraged us to lead drills, and after a while we were on autopilot and ran our own drills while the captain just sat back and smiled. Looking back, I think he pulled a "Tom Sawyer" on us.
- **Leadership.** Although our company was like a big, happy family, there never was a question of who was in charge. Both in the firehouse and on the fireground, Captain Burns was the boss. He had our respect, and when he spoke, we listened.
- **Lead by example.** Captain Burns never asked us to do anything that he wouldn't do himself. He led us into battle and was always in position to supervise our actions.
- **Monitoring safety.** Sometimes we had an overly aggressive group of young firefighters on Squad 2. Big Guy kept us in check without deflating our spirit. He let us get cocky—just not so overconfident that we jeopardized safety.
- **A student of the job.** Setting an example for us, Captain Burns received his Masters Degree in Public Administration. He always had a book in his hand and encouraged us to do the same. We held company critiques regularly, with a goal of continuing to learn. This helped transform us into thinking firefighters.

Captain Burns's motto was "Bring 'em all home"—and he did. He always watched out for his firefighters. On some occasions he single-handedly saved the lives of his crew, as you will read in Story 24.

> *He always watched out for his firefighters.*

He also turned out leaders. Every firefighter from our crew has been promoted over the years, and among this group are six Chief Officers. Needless to say, the fire service needs more officers like Bill Burns.

Lessons Learned

- **If you want to be an outstanding fire company officer,** you should follow the example set by someone like Bill Burns.

- **Firefighters can become leaders themselves** by identifying and practicing the traits of a true leader.

Discussion Questions

1. What are some characteristics of great leaders?

2. What are some of the ways a firefighter can prepare to become a leader in the fire service?

3. What is meant by "pulled a 'Tom Sawyer' on us"?

4. Are personal and company goals important? If so, why?

5. What is the importance of a mentor? Does a mentor have to be the firefighter's officer?

When Duty Calls

Firefighter James J. Regan
37 Years as Fire
Protection Engineer
8 Years as Volunteer Firefighter

In this business, you just never know. The unexpected is what challenges us to be on top of our game. That Sunday was cold and sunny, typical for December. It also was my teenage son Jim's 15th birthday. I had been a Fire Protection Engineer for more than twenty years at the time, and a fire buff my whole life. Shortly after noon I heard on my scanner that the Department was battling a 2-11 alarm. I grabbed my scanner and camera, and my son—an aspiring firefighter—and I left our suburban home, heading for the city.

Invitation to a Fire

The fire, in the interior of a four-story corner apartment building over storefronts, was running the walls and interior plumbing walls. From the perspective of a fire buff and amateur photographer, the fire was unspectacular, just a tough interior firefighting job. I shot a few pictures, then Jimmy and I headed for Squad 2 on the northside.

Who knows—maybe we would catch a "hit" with the Squad. This heavy rescue unit covered working fires in the northside of the city. We arrived during the third shift, and Captain Bill Burns, a/k/a Big Guy, was in command. We had been friends for more than eight years by then. The firefighters on the squad that day were Lenny, Danny, Tim and a few others that I didn't know. Jimmy and I no sooner walked into the kitchen than the speaker crackled, sending Squad 2 to a working fire. I asked the Big Guy if we could come along, and he waved us on.

As we pulled out of quarters, we could hear the bells ringing. The Fire Alarm Office was striking a box (2nd alarm)—a sure sign of multiple calls reporting a fire. We rushed south past Engine 76's house and followed the engine company into the 13th Battalion. As the guys were getting out of the crew cab, I looked past them at the large building and warned, "Be careful!

They approached the rundown, U-shaped, courtyard building. Steam and light gray smoke were coming out of first and second floor windows near the rear. From the looks of things, the fire companies were getting a handle on it. Most of the fire appeared to be in the rear of the building, so Jimmy and I headed to the rear alley, my camera and portable scanner in hand. Across the alley to the south was a large vacant lot, so we had a lot of room to observe and take a few pictures.

Flames danced high over the firefighters' heads as they scrambled for their aerial ladder.

While heavy smoke was pouring out of several windows, there was no visible fire. A truck company was busy ventilating the roof over the rear. We could hear the screams of the K-12 saw as it chewed through the roof deck. Soon heavy black smoke poured from the roof opening, and within seconds the smoke lit up. Flames danced high over the firefighters' heads as they scrambled for their aerial ladder.

The large hole cut in the roof had changed the conditions inside the building—and not for the better. Flames were pumping out of the windows on the second floor now, and the third-floor windows were all starting to light up. My scanner had an extension speaker similar to a real Handi Talkie™, so I could hear the fireground action while using the camera. As soon as the windows lit up, the radio traffic picked up, with company officers calling for hoselines and checking on their crews.

Suddenly I heard a muffled message amidst all the traffic: "Squad 2, we're trapped!" The message was repeated. I looked up and down the alley to see if anyone else had heard that message. Apparently no one had.

Big Guy in Action

Inside the building, things were getting tense.

Inside the building, things were getting tense. Upon arriving, the captain had split the company into two teams. Tim, one of his senior guys led one team, assigned to perform a primary search on the second floor. Big Guy took Danny and Lenny with him to search the third floor. The building had been divided into single-room occupancies or SROs, and the occupants shared one bathroom down a long hallway. These fires are always tough fires because there are many occupants, a lot of locked doors, and long hallways.

The Captain's team had been forcing in doors and performing a zero-visibility primary search. The smoke was getting blacker and hotter and was moving faster. Big Guy recognized these as danger signs that the smoke was about to light up. Firefighters from the search team, now bellied to the floor, tried to make it to the stairway they had come up. As they crawled closer to the stairway, the heat was even more intense. Suddenly—flashover! Everything lit up at once, and the heat drove them back down the hall away from the stairs.

> *Suddenly—flashover! Everything lit up at once.*

Big Guy found a door that they had forced open only minutes before. "In here!" he yelled to the two firefighters behind him. They all rolled into the room, and the captain slammed the door shut behind them. This barrier would buy them a little time.

The captain had to throw his back against the door to hold it shut because the lock had been destroyed when the door was forced during the search. He called in a few messages of firefighters trapped, but with the heavy radio traffic, he didn't know if he had been heard. "Get someone's attention, and get us a ladder to the window," he ordered his men.

They took out the window and tried to attract the attention of firefighters in the distance. The door was getting harder to hold, even for the Big Guy, at 6′6″ and 280 pounds of pure muscle. The pressure from the flashover was incredible. He was holding the door with every ounce of strength he had while flames were pushing over his helmet like a blowtorch as the top of the door burned away.

Back Outside

Danny appeared in a third floor window, trying to get anyone's attention. I got off a quick shot on my camera, and then another as Lenny appeared momentarily. Apparently, nobody besides Jimmy and I had heard the message from the captain. I gave my camera to my son and told him to come with me as I ran down the alley looking for help.

When we turned the corner, Fire Commissioner Ray Orozco, Sr., was walking toward the rear. I told him, "Squad 2 is trapped on the third floor. We need to get a ladder down the alley."

He called for help on his radio as the three of us ran for a truck parked in the street. Somehow we managed to take a 38-foot extension ladder off the truck. We half-carried it and half-dragged it down the alley across the ice. When we approached the area, firefighters were directing a stream over the Squad's heads while black smoke poured out. Firefighters grabbed the ladder from us and quickly raised it to my trapped friends.

Lenny called for a ladder while Captain Burns held the flashover behind the door.

To the Rescue

Danny yelled to Big Guy, "We've got a ladder. Let's go, Cap!" The captain, now with heavy flames blowing across the top of his helmet, said, "I will, but you two get out first." Lenny dived out the window onto the waiting ladder. Danny called for the captain again.

"You go first," the captain yelled. Once Danny was safe on the ladder, Big Guy wedged a dresser up against the door and dived for the window. The dresser was no match for the pressure, and flames followed the Big Guy out the window—too closely.

Safely on the ground, they looked at the captain's plastic helmet with a large dimple melted in the side, and the Bourke® (or) ™ eye shield had curled up from the intense heat. It was a close call, and they all knew it. The captain stood there for a while with his melted helmet in hand, looking at the flames roll out of the third floor window. The dimple in

> *It was a close call, and they all knew it.*

his helmet was large enough to hold a baseball and the helmet was almost completely burned through.

Now, fifteen years later, my son Jim is a firefighter and I'm a volunteer firefighter in a suburban Department. I'm a little heavier and a lot more philosophical, people say. We still get to the city and visit Squad 2, even though the entire crew has changed. I don't know if we made a difference that cold Sunday. I just hope someone is listening if Jim ever needs a ladder!

Bill Burns's melted helmet.

Lessons Learned

● **Good communication is necessary for safety.** Reports after the deaths of firefighters in the line of duty usually have some common denominators—among them, a breakdown in communication and lack of accountability. These two go hand-in-hand. You can't have one without the other.

● **A Mayday procedure must be in place.** The word "Mayday" frees up the fireground radio channel for only the trapped members and the RIT (Rapid Intervention Team) companies.

● **Training, training, training!** We must continually train to keep firefighters safe. Every ounce of sweat that we shed in training keeps us safer in battle. We must drill on how to react in emergency situations so our reaction becomes second nature.

Discussion Questions

1. What are some of the warning signs of flashover?

2. In forcible-entry classes, firefighters learn the importance of controlling a door and maintaining the integrity of the door so it may be used as a barrier. What are some methods of controlling a door?

3. Why is a secondary means of egress important?

Mayday! Firefighters Down!

The day started out sunny, with 40-degree temperatures. "We still have eighteen hours to go on this shift, but it's quiet," I remarked to John my driver. As captain, I was a little nervous because he was the only regular firefighter assigned to the Squad that day. Several of the guys had off-days, and four were detailed to us just for the day. Then, at 2:00 p.m., Squad 2 was dispatched to a fire punctuated by black smoke rising into the blue skies in the distance.

Captain Patrick Maloney
23 Years in the Fire Service

"Give Me a Search"

On the radio we heard the first company's description as we hustled to the fire. It was a three-story, U-shaped, courtyard apartment building, ordinary construction, with fire on all three floors. The 9th Battalion Chief called for more help and pulled a box alarm. As we turned down the block, the Chief called me on the radio.

By the time we got out of the rig, the Chief was calling again. "There's a report of people trapped in the exposure building to the south. Give me a search."

"Message received," I replied.

With the knowledge that people were trapped inside, we quickly sized up the exposure building while we hurried toward the front door. It was what is known as a "three-flat building," probably 1930s ordinary construction. Flames shot out of the windows of the courtyard building like a giant blowtorch. With only a narrow gangway between the two buildings, there was a good chance we'd find fire in the three-flat. The smoke was already getting heavy, and we masked up as we reached the first-floor landing.

At our morning roll call, I had split our company into three teams, and each team had a radio. I assigned one team to check the first floor, and one to check the second floor. My partner Mike, who was one of the detailed firefighters, and I would check the third floor.

The first-floor door was no match for my firefighters, and with one kick they were inside. The conditions were light smoke and maybe 10-foot visibility. Staying low in a duck walk, they slipped inside. The same scene played out on the second floor, except the smoke was thicker and hotter. I watched as my second team disappeared into the smoke. When I called the Chief for a hoseline to protect our stairway, he told me one was already on the way. My partner and I reached the third-floor landing, forced entry, and slid into the apartment in a low crawl. It felt like there was a lot of fire below us, but my second-floor team hadn't reported any fire.

> *I watched as my second team disappeared into the smoke.*

We crawled quickly through the two-bedroom apartment, doing a right-handed search. By the time we reached the first bedroom, the smoke was so thick that I couldn't see the bright sunlight through the window unless my face was against the glass. I didn't want to open any windows until we had a hoseline, because it was getting hot and I thought the whole place would light up if I did. I instructed Mike to stay at the door while I searched the last bedroom. The floor felt spongy to me, but this didn't fit with what I knew—not with ordinary construction and only minimal fire below us. But I couldn't ignore the fact that sponginess is a warning sign of collapse.

After checking with my teams, I relayed to the Chief that the primary search was complete on all three floors. My partner and I crawled back to the front door, with conditions getting worse by the second. An engine company brought a 1¾-inch hoseline up the front stairs, but it came up short, reaching only to the front of the hallway. Just then I heard firefighters with a hoseline in the back of the apartment. "Oh, oh!" I thought, "the spongy floor!"

Quickly I crawled to the back of the apartment, and I bumped into a firefighter with a hose in his hand outside of the rear bedroom. Grabbing him, I asked, "Who do I have here?"

"Engine 78."

"Don't go in there," I warned him. "The floor is bad."

By that time I realized that the hallway floor was also getting spongy. There must be a lot of fire below us. "Back out!" I yelled. "Back out now!"

The heat was getting unbearable back there. I started crawling toward the front door and grabbed my radio: "We're backing out of Three! Watch out on the second floor!"

Escape

It happened quickly: Everything went black, I felt the floor dropping, and then I heard a loud "boom!" To escape down the hallway, I had to bear-crawl up a steep slope. Flames were chasing me amidst the screams of firefighters in the darkness: "Help! I'm burning!"

I grabbed the radio: "Mayday! Mayday! Mayday! Firefighters down on the third floor! Sector three! We need a line on two, and ladders to three!" I started pushing the firefighter on the hoseline out the front door onto the stairs. "Get that line down to the second floor," I ordered, "78 must be hanging over the fire."

"They have a line down there already, Cap," he replied. Confused, I started back toward the screams.

A Safety Chief who had just arrived on the scene came running up to me. "I know where they are," I yelled. He followed me back down the weakened hallway. I said to the guy on the nozzle of the short line, "Keep it over our head." The hallway floor felt like it would drop in any second, but these were our brothers, and we had to save them.

In the blackness I finally reached a downed firefighter and grabbed him. "Not me," he said. "Grab him." He was out of air and choking on the thick smoke, but heroically he was lying there, holding onto another firefighter, keeping him from falling down a steep incline.

> *In the blackness I finally reached a downed firefighter and grabbed him.*

"Grab this one," I yelled to the Chief, "I have another one." The Safety Chief grabbed the gasping firefighter and started down the hallway.

I grabbed the second downed firefighter by the shoulder strap of his SCBA harness. I tugged with all I had, and he didn't even budge. Flames were licking at his legs, and he wasn't screaming any more. The floor was going to fail soon. I had to get him. I tugged harder the second time and felt a slight movement. I couldn't worry about the floor beneath me. I had to get him. Grabbing his harness with both hands, I tugged with all I had, and he came up the slope.

Now I was out of air, my face piece against my lips. I tore it off and started sucking for air off the carpet, choking and gagging as I dragged him down the hallway toward the hoseline. The Safety Chief came back down the hallway and helped drag him to the stairs. With the help of other firefighters, I brought him down the front stairs.

"Squad 2 to 223—we have him," I yelled into the radio. "Get an ambo to the front of the building! It's Battalion 9!"

Out of the smoke, I could see who it was. It was Leo, the still alarm Battalion Chief. Leo had come in the back door to get his companies out just before the collapse. Now he lay there unconscious. Leo was one of the most respected chiefs on our job, and I thought we had lost him.

Taking Stock

Once Leo was delivered to the ambulance company, a few squad members and I changed out the fresh-air bottles and started back up. On the radio we could hear the Incident Commander taking roll, and several companies were missing members. As it turned out, all the firefighters had made it out on their own. Accountability went awry because the companies were split as they exited opposite doorways. The excitement of the Mayday and the roll call added too much radio traffic.

While we were bringing Leo to the ambulance, the Safety Chief and another firefighter had to escape from the back bedroom by kicking through the back wall and sliding onto a lower roof. Their narrow escape was a result of all the confusion. They were sure that other firefighters were trapped because they had found a boot and a helmet at the collapse hole. Also, several firefighters were unaccounted for during the roll call, which added to the confusion.

After things were brought under control, I discovered something that all of us had overlooked: In the quick size-up of the building, we hadn't noticed that the third floor was an addition. About 5 feet above the existing roof of the original two floors, the sidewalls were bridged with lightweight laminated I-beams. Then it was decked with plywood before construction of the new third floor. It didn't look like a new addition, and I never would have expected lightweight construction in a 1930s three-flat. The lightweight construction explained the spongy feel and early failure of the floor on the third floor. The confusion was compounded because most of the fire was burning below us in the newly formed cockloft.

Leo spent several days in Intensive Care and several more in the Burn Unit. Another firefighter, the lieutenant of Engine 78, also received bad burns on his legs. The good news is both have recovered well, and the rest of us received lessons we will never forget.

Lessons Learned

● **We must take the extra second to read the building.** The lightweight construction was overlooked in this instance, and it will fail in a fraction of the time that ordinary construction takes to collapse. New construction, rehab work, and additions must be considered lightweight until we can rule this out.

● **A secondary means of egress is essential.** We must know where we are in the structure at all times and have at least two ways out in an emergency. The way we entered the building is not always the best way out.

● **We must trust the instincts that we develop from our experiences and training.** When something tells us it's not right, it's usually not right!

● **During Mayday emergencies, radio traffic has to be cleared.** When a firefighter is down, this must take precedence over all other communications. Only companies involved in the rescue, such as RIT teams and the Incident Commander, should be allowed to use that channel. Others must switch channels, or if this is not an option in your department, cut down on needless radio traffic.

● **We must stay with our assigned partner.** We have to know where our partner is at all times and never freelance.

● **Accountability is essential.** Officers have to know where their people are at all times. Firefighters who are split up into teams must report to their officer face-to-face or via radio immediately when an emergency arises. Rescuers have died trying to save firefighters who were already safe.

Discussion Questions

1. What is the probability of collapse of lightweight construction in a structural fire, and what warning signs can we expect?

2. How can communication breakdown lead to accountability problems? How does this affect firefighter safety?

3. In the fire department in your locale, what is the roll call accountability check procedure after a collapse?

Physical Fitness

Firefighter Darryl Johnson
15 Years in the Fire Service

When I came on the Department, I thought I was in good shape. Before I entered the Department, I was a competitive body-builder. I was as big as a house and felt strong as an ox. People would tell me that I looked like I was made for this job, and I started to believe them. I kept lifting heavy and eating heavy. Cardio workouts were unheard of for me. When I graduated from the Fire Academy, I was assigned to Engine 107, in the shadow of the housing projects. These were high-rise, low-income buildings where the elevators routinely did not work—a bad omen.

Tested Under Fire

One day we responded to a fire on the eighth floor. From the street we could see large puffs of dark gray smoke coming off the eighth-floor balcony. As our company started up the stairs, I was wearing all of my fire gear—a Scott Air-Pak on my back, 100 feet of 2½-inch hose on my shoulder, and a high-rise bag of fittings in my hand. When we reached the fourth floor, one of our guys couldn't go any farther without resting. My officer knew that time was too critical for us to stop. He ordered me to take the other firefighter's hosepack, and we continued up.

The addition of another 100 feet of 1¾-inch hose was like the proverbial straw that broke the camel's back. By the time we reached the eighth floor, sweat was pouring off of my body and my heart was pounding out of my chest. As soon as we got there, I dropped the equipment on the landing and I dropped to a knee to catch my breath.

> *Sweat was pouring off of my body and my heart was pounding out of my chest.*

"What are you doing?" my officer asked, and without waiting for a response, added, "Now it's time to go to work."

I don't know how I made it through that fire—on pure adrenalin, I imagine. When we returned to the fire station, my heart was still pounding. "Never again," I told myself.

From that experience I realized the importance of physical fitness on this job—and I don't mean just looking good. There are "show muscles" (muscles that make a person look good), and there are "go muscles" (muscles that affect job performance). I had show muscles already, but I needed some go muscles. It was back to the drawing board.

Developing Go Muscles

I used the same drive and determination that made me a good bodybuilder to become a fit firefighter. I studied the needs of a firefighter and tailored my workouts to meet these needs. One of my greatest needs was stair training, which provides a tremendous cardio workout. I lived in a high-rise apartment building, with stairs just outside my door. Climbing the stairs gave me a superb workout, and an emotional boost besides.

Then I started mixing calisthenics with my weightlifting routine, which improved my flexibility and stamina. Immediately I noticed an improvement in my firefighting abilities. I started designing exercises that would better meet the needs of firefighters, and this inspired an interest in becoming a personal trainer.

A New Role

I took a 40-hour personal trainer course and received my state certification. This further piqued my interest, and I continued my education in the study of kinesiology, the science of motion. I continued to take classes and at the same time worked part-time as a personal trainer.

When I started working with other firefighters, the word spread. In the summer of 2000, I was asked to be a physical fitness instructor at the Fire Academy. I enjoy the challenges and the rewards that come with this position. The Academy offers two six-month classes each year, with a typical class size of approximately 100 candidates. These new recruits come in a wide range of ages, sizes, and physical conditions.

For each new class, I first talk about the physical demands on fire-fighters and the dangers of the job. I introduce the safety issues related to physical fitness, and stress-related injuries and deaths, noting that cardiac arrest is the leading cause of firefighter deaths on the fireground. Then we talk about how to modify their lifestyle to prevent injuries and to make them more effective firefighters. I explain my goals and expectations for them, and the standards they must meet.

We start every new class with a PFT (physical fitness test). This test consists of charting the height and weight of the candidates, along with their time in a 1½-mile run, number of push-ups and sit-ups in two minutes, and total number of pull-ups. Based on these results, we chart a course, and a monthly PFT monitors individual progress. We set reasonable and attainable goals, and we strive to develop teamwork, motivation, and mental toughness— three qualities needed to succeed in the fire service.

> *We strive to develop teamwork, motivation, and mental toughness.*

Because I believe in leading by example, I don't ask any of the students to do what I am not willing to do myself. This sends a positive message to the candidates. Every other day, we run to cadence in formation. On the other days we do calisthenics together on a cadence count. We don't start off by just doing a new exercise. First we teach the exercise and explain what it is designed to accomplish, emphasizing form, discipline, and safety.

The Result

Ninety-nine percent of a successful exercise routine is mental, and the mind wants to quit long before the body has to stop. Mental toughness helps push us through these barriers. This same mental toughness will later help push these firefighters down tough hallways.

Our program has met with great success. A class is only as good as

> *A class is only as good as its trainer.*

its trainer. Good trainers never rest on their laurels and must strive to stay current in their education. I just finished a 40-hour peer fitness trainer course, and I'm always looking for more education. Recently I was named Wellness Coordinator for the Fire Department. We also have a weekly exercise segment on one of the major television networks.

Remember that firefighter who couldn't run up to the eighth floor? Since then, I've run four 26-mile marathons and a 31-mile ultra marathon over rough terrain. If I can do it, other firefighters can, too!.

Lessons Learned

- **Physical conditioning is important in fighting fires.** The very nature of our job dictates that we go from zero to 60 in six seconds. It can save your life, prevent injuries, and improve the quality of your life.

- **Physical fitness can prevent injury and death.** The most common cause of death on the fireground is cardiac arrest, and the most common injury is to the lower back. Fitness will help prevent these outcomes.

- **We should strive to be the best we can be.** That includes becoming as physically fit as possible.

- **Mental toughness goes hand in hand with physical toughness.** Together they propel us to accomplish our goals.

Discussion Questions

1. What modifications can you make in your lifestyle to improve your overall health?

2. What are the benefits of running and exercising to cadence?

3. How can firefighters improve their physical conditioning and promote company cohesion at the same time?

Whatever Can Go Wrong . . .

Firefighter Raymond
E. Cullar
18 Years in the Fire Service

That summer day in 2002 was hot and sticky. We had just finished dinner and were sitting around when the alarm came in: "Engine 62, Truck 27, and Battalion 22 respond to a still alarm. . . ."

Fire in an Abandoned Building

With the crackle of the speaker, eleven fire-fighters scrambled to their rigs. Rush hour had just ended, so traffic wasn't too bad. I was the driver, and within a few minutes we reached the fire building, a 2½-story, aban-doned brick building with plywood covering the windows. Fire was blowing through the pitched roof in the rear.

I asked my captain, "How could it be burning like this so early in the day?"

"I don't know, Ray. You'd think someone would have smelled it before it started to burn through the roof."

At morning roll call, my partner and I had been assigned to do roof ventilation, but the fire had already vented itself. Instead, the captain had us come inside with him. Sinking the pick of his axe into a corner of the plywood, one of our firefighters removed the boards covering the front door. Thick, dark gray smoke started to roll out.

"Send the water!" the engine officer yelled.

The line jumped as it filled with water, and the nozzle hissed as a fire-fighter bled off the air. Soon water, first in a sputter and then a large volume, crashed against the wall. The firefighter quickly shut off the flow, and we were ready.

"Let's go get it," the captain ordered, and we slipped inside the smoke-filled building.

Our truck company started popping the plywood off the windows, and the engine company progressed to the inside stairway. As we tried to advance, heavy black smoke rolled down the stairs on top of us. Why wasn't the heat lifting? It was so intense that the engine company couldn't advance up the stairs. I'll bet the fire hasn't burned through the roof yet. It probably lapped out of the back window and ignited the roofing, I thought. That's why this heat isn't lifting. I shouted for my partner but couldn't find him in all the confusion.

> *Heavy black smoke rolled down the stairs on top of us.*

I have to open that roof, I decided, and went outside, figuring someone from the second truck could help me. Nobody was outside except an engineer who was busy on the pump panel, so I jumped up on the turntable of the truck and raised the aerial ladder to the peak.

The building had a 12/12 pitched (very steep), gable (two-sided) roof. Grabbing my axe, I hustled up the ladder without my partner. I straddled the ridge and worked my way toward the rear. I had to swing around two chimneys to reach the fire area. As I got in position to open a hole, I could feel a lot of heat below me. Suddenly there was movement in the roof deck and I knew it was going to let go. Sensing a roof collapse at any second, I headed for my ladder.

"Crash!" The roof started falling. I dived toward the chimney. When the roof collapsed, the chimney separated from its metal flashing and I barely managed to sink my fingertips into the flashing. Flames now were rising over my head. I was hanging by my fingertips on a steep roof, 50 feet up in the air, to a piece of sheet metal the thickness of a postcard. To make matters worse, the metal was barely fastened to the roof deck, and nobody knew my whereabouts. I was afraid to move a muscle, as one slight move could dislodge the flashing and I would fall to the ground.

> *"Crash!" The roof started falling.*

The alarm had escalated to a box alarm (second alarm). I lay perfectly still and slowly reached for my radio with my free hand to call for help. I grabbed the microphone only to realize that the battery was dead. Could things get any worse?

Unexpected Help

I yelled for help, but nobody could hear me. It was like a bad dream. As I clung for my life by the fingertips of one hand I could see several people and yet nobody seemed to notice me. I couldn't hold on much longer! Was the fall going to kill me? Finally I looked down and saw a woman in a neighboring sideyard looking up at me. I hoped she knew that I was in trouble. I watched as she then ran to the front of the building, screaming at one of our chiefs and pointing towards me.

He looked up to see what she was pointing at, and then turned away. "Oh, no!" I surmised. He didn't get the message.

Like they say, whatever can go wrong will go wrong. The roof deck was getting hotter, and I knew it would light up any second now. That meant I had to get out of this mess myself. I had to try something or surely I would die! Grabbing the flashing with both hands, I slowly pulled myself up, silently breathing, "Please don't break off!" Slowly I raised my upper body toward the weakened ridge and with one motion pulled upward. The flashing broke off—I lunged forward—throwing my hands up towards the ridge of the roof. My body started to slide down the roof and suddenly I stopped. I wasn't falling—my fingertips now clung to the ridge—I made it! Somehow I managed to pull my body onto the ridge and then scurried back to my ladder.

As I was climbing down, the Chief looked up at me in disbelief. He had been certain that nobody was on that dangerous roof. "What were you doing up there? Where's your partner?"

I didn't have an answer. I had gotten caught up in the situation and knew I had messed up. I had a close call that day—too close—but I learned some valuable lessons that will stay with me throughout my career.

Lessons Learned

- **Freelancing is not a good idea.** Firefighters shouldn't separate from their partners. In this story, it meant wandering off to perform a task other than the one assigned. This lack of accountability can be deadly on the fireground.

- **Breakdowns in communications can be disastrous.** This includes verbal communications and radio communications, and this story demonstrates a breakdown in both areas of communication.

- **Radios can save lives.** We should check our battery throughout the tour of duty and change it before a problem arises.

Discussion Questions

1. Why is freelancing dangerous on the fireground?

2. What do you think caused the heat to push down the stairway in this story?

3. What is the relationship between communication and accountability?

Communication and Accountability

Pat McAuliff
28 Years in the
Fire Service

The worst nightmare in the Fire Service is to lose a brother or sister in the line of duty. In the United States more than 100 firefighters die in the line of duty each year. The most we can do is to try to learn from each fatality and teach others the lessons we have learned. The post-incident reports often reflect a breakdown in communication and accountability. Failure in these two areas dates back to the days of the bucket brigade, yet we sometimes make the same mistakes today.

A Fire on Campus

One of my own early lessons in communication and accountability happened on a sunny summer day in 1979. Our three-man company was the second engine to respond to a fire on a college campus. At around 6:00 p.m. we arrived at a six-story building containing classrooms and laboratories. Firefighters from the first engine already had laid a supply line into the standpipe system and were inside investigating the location of the fire.

A few firefighters came into the lot with red lights flashing on the dash of their personal cars. At that time, both paid and paid on-call firefighters staffed our fire department. Off-duty firefighters often just showed up at the scene, grabbed an SCBA, and followed the hoseline to the fire.

We spotted our apparatus in the front of the building, grabbed hose-packs and some extra air bottles for the first engine, and started to go in. A small group of people in front of the building informed us that they evacuated the building when the alarm went off. They were fairly sure that nobody else was inside.

The light from large skylights at the top of the atrium was obscured by the smoke.

As we entered the six-story atrium, it was quickly filling up with black smoke. The light from large skylights at the top of the atrium was obscured by the smoke. We weren't concerned that occupants were still inside, because of the information we had received from the people outside, and because we knew that the fire alarms had been activated. We concentrated on helping the first engine company find and extinguish the fire.

First we placed the spare air bottles to the side of the stairway, in an easy-to-find location. Masked up, with hosepacks over our shoulders, we started up the stairs. Because the heaviest smoke was on the third floor, this seemed to be where the fire was located. Firefighters from the first engine were on this floor, having led their hoseline down a long hallway. Now they were retreating to replenish their air supply. We told them where the extra bottles were staged, and we crawled up to take their place on the line. Thick black smoke was rolling down the hallway at us, and the heat was increasing every inch we crawled forward.

Our destination was somewhere out there in the blackness.

We made our way down the long hallway, our destination was somewhere out there in the blackness. After a few minutes we finally saw it—a faint glow at first, and then flames billowing out of a room down the hall to our left.

"Ringggg." My partner's low-pressure alarm went off.

Within a few seconds, my alarm also started to ring. We had taken too long to locate the fire, and now we had only five minutes to exit the building.

"Let's go!" our Officer yelled, his voice muffled by his face mask. The first engine traded positions with us once again, and we headed back down the stairway. By the time we reached the lobby, our masks were clinging to our faces as we consumed the last of the air.

"Where's Rick?"

We quickly donned our SCBA harness and were changing out our air bottles when someone came up and asked, "Where's Rick?" He hadn't ridden the engine with us, so he must have come on his own.

Because he was so faithful in responding to fires, our officer concluded, "He has to be here somewhere." A call went over the fire radio to all companies working on the scene.

Nobody knew of Rick's whereabouts. We had never dealt with a missing firefighter before, and we started to panic. Looking up at the smoke pouring out of that six-story building, we asked ourselves, "Where do we start?"

Ordering everyone out of the building, the Chief called the roll while the fire continued to burn. The first engine officer firmly believed that the fire could be extinguished if he were to take a crew back into the building immediately, but the first concern was for the missing brother: "What about Rick?"

Pat McAuliff (left) and team were changing their air bottles when somebody asked, "Where's Rick?"

The Chief sent the crews back into the building, some to fight the fire and others to search for the missing firefighter. At the door, the Fire Marshal had a notebook and took down the name of every firefighter who entered the building, then crossed out the name when he came out. Looking back, that notebook and pencil constituted the Flintstones' version of the modern-day fireground accountability system.

While others fought the fire, we started a thorough top-to-bottom search of the entire building. As a team, we followed walls to navigate in zero-visibility conditions. Finally the first engine was able to bring the fire under control, and we were able to accomplish a more thorough secondary search. For close to an hour, our frantic search continued. Each time an air bottle was consumed, we exited, quickly changed out a bottle, and returned to the search. The Chief called for extra air bottles to be shuttled to the scene.

> *For close to an hour, our frantic search continued.*

At some point we began to question, "Is Rick really here at all?" "Did someone actually see him, or did we simply think he was joining up with them inside the building?"

At about this time, a dispatcher radioed the Chief that he received a phone call from a paid on-call firefighter who lived in a nearby town. It was Rick. He never was at the scene. In fact, he had gone home after the call had been dispatched, missing the general alarm that asked all firefighters to respond. He just happened to turn on a scanner, heard all the radio traffic, and realized that he was the one they were looking for. Firefighters had been needlessly risking their lives for him while he was safe and sound several miles away. We received a valuable lesson on communications and accountability that day.

I believe that a big problem in the fire service is that we sometimes do not share our mistakes. Maybe this is out of fear of admitting failure, or fear of being ridiculed. The fire service is filled with some of the most intelligent people I have ever met, yet sometimes we are slow learners. Why are we still making the same mistakes? Why is accountability still an issue today? Let us all be committed to never letting a firefighter's death go unnoticed. Let's continue to learn from every incident, and never let our egos get in the way.

Air packs are changed after firefighters descended from the third- story blaze. Staff Photo by Steve Cast

Firemen fight blaze for 2 hours

From page 1A, col. 6.

p.m. when smoke detectors sent an alarm to the A&M police department which notified firefighters.

Mary Corbett, a technician, had been working in a room next to the lab. "I heard the alarms and smelled the smoke all in the atrium and got out," she said. About 10 persons, mostly custodial workers, had evacuated the building when firemen arrived.

The $10.8 million building has a common atrium connecting two five-story concrete structures. "The way this building is designed, it keeps a fire from spreading, but it also creates an oven," College Station Fire Chief Douglas Landua said.

"The building is theoritically not burnable, but the contents are," Kirk Brown, fire marshal for the building and an associate professor, said. "There's every chemical you want to name in there and they are explosive," he added.

Fire inspectors began to rummage through the lab to determine the cause of the blaze this morning.

Plastic materials used in experiments caused much of the smoke and fueled the fire, officials said. "The things in there are very valuable, very expensive and very dangerous," Brown said.

"But these are the things we have to have to stay ahead in the field of research and academic advancement," he added.

College Station firefighters take a breather after assault on blaze. Staff Photo by Steve Cast

Pat (bottom right) and crew after the two-hour firefight.

Lessons Learned

- **We must account for all of our firefighters at all times.** We can't wait until someone asks, "Has anyone seen Rick?" Officers must know where each firefighter is at all times. Freelancing should never be allowed.

- **Good communications is the key to accountability.** Firefighters report to their officers in person or via radio, and officers report to the Incident Commander or Sector Chief.

- **We must utilize a Mayday and RIT or FAST System.** For a plan to work in emergency situations, we must practice. RIT/FAST companies must take their jobs seriously. We would rather be fighting the fire than standing by, but we must be disciplined. Most of the time we will just stage our equipment, size up the building and conditions, and then stand by, but when we are called upon, it will be for the most important duty we will ever perform—to save the life of a brother or sister firefighter.

Discussion Questions

1. Is lack of accountability still an issue today? If so, what can we do about it?

2. What modern-day equipment and procedures not available at the time of this story would be useful in missing firefighter situations today?

3. What actions does your Fire Department take when informed that a firefighter is missing?

Growing Up on the Job

Lieutenant John Gariti
20 Years in the Fire Service

Igrew up on this job—literally. My father, Ben Gariti, was a Chief in the Fire Department. I idolized him, and all I ever wanted to do was make him proud. I entered the Department at age twenty-one. The fire service has been a tremendous learning experience for me, and over the past twenty years I have matured physically, and mentally. One incident particularly stands out in my mind as a lesson learned that helped me grow into the fire company officer that I am today.

Hot Dogging

After only a few years on the job, I was detailed for the day to a busy downtown company, Rescue Squad 1. About 1:00 a.m. on a balmy summer morning, I was in bed after a busy day. The call came in: "Squad 1, take in a jumper on the expressway." We knew that a jumper is someone threatening suicide, and these incidents often prove to be challenging.

Seconds later we were running for the rig and within minutes we were pulling up on the scene. A woman was standing on a narrow steel platform below a sign, which was attached to a bridge about 25 feet above the expressway traffic.

A Battalion Chief was leaning over the bridge trying to convince her not to jump. He had called for State Police to stop the expressway traffic, a tower ladder, and an air bag to stage under the bridge once the traffic was stopped. Once the air bag could be safely deployed it would be our safety net for both the victim and us rescuers. Our Squad officer instructed a few of the firefighters to start suiting up in three point harnesses for a potential vertical rescue attempt.

> *I had other ideas, and I saw my opening.*

I had other ideas, and I saw my opening. The jumper was being distracted by the Chief and was looking in his direction. I lowered myself about 6 feet onto the narrow, foot-wide platform and started crawling up behind her. Traffic was still moving below me as I inched toward her.

When I approached, the Chief's eyes widened and locked on mine.

The jumper kept looking in his direction as I pounced on her from behind. I grabbed her and pinned her to the platform, where a struggle ensued.

"You're a pretty lady and have a lot to live for," I offered.

A manly voice answered, "I'm not a lady."

Oh, oh.

He added, "I have HIV, and I don't have any reason to live." Without warning, he started to struggle violently. All I could do was pin him on our narrow perch and hold on for my life. He was now bucking like a wild bronco. My arms wrapped around him and I clung to the metal platform with all my might. He was trying to kill us both.

On the Ground

Sirens were suddenly screaming below us, and lights were flashing everywhere. The expressway traffic below us came screeching to a halt. The State Police had quickly stopped traffic below us. Another firefighter from our squad dove out onto the perch and helped pin the wild man to the platform.

Within a few minutes the tower ladder was positioned below us, its basket directly below me. A firefighter in the basket grabbed the legs of the jumper, and pulled him into the basket. The other firefighter and I each grabbed an arm and forced him down to the floor of the basket. We lowered ourselves into the basket to hold him down. We made it—he was safe. I looked up on the bridge for approval.

The Chiefs eyes seemed to stare right through me. He said nothing, but he sure didn't look happy. Standing next to the Chief, my officer was waving his arms wildly and yelling at me, "What's the matter with you? What were you *thinking*?"

After the basket was lowered to the expressway, we passed the man over to some waiting police officers. They handcuffed him for his own protection, and he was transported for psychiatric evaluation.

After cuffing our jumper, one of the State Police officers congratulated me: "Nice job. You should get an award for this."

After seeing the faces of the Chief and my officers, I knew better. I shook my head and said, "I think it's more likely that I'll get fired."

When we got back to quarters, the officer took me into his office and read me the riot act. He accused me of being a "hot dog" and said that I'd never be allowed back on his squad. He told me that I had jeopardized the safety of every firefighter on the scene, and I was lucky he didn't file charges on me.

After my officer cooled down, he said I reminded him of someone he knew many years ago. He was talking about himself. His father also was a Chief when he had entered the Fire Department, and he also felt he had something to prove. He wanted to make his father proud too. That guy grew up though. He realized that the way to make his father proud was to be a safety-conscious leader, and not a reckless hot dog. That night on the expressway, I grew up too.

My officer that night went on to become a Chief on our job. He has called many times over the years to recruit me to work with him. We remain friends to this day.

Lessons Learned

- **Freelancing is dangerous.** This Lesson Learned is obvious here, as an overly aggressive freelancer jeopardized everyone's safety, including his own.

- **To operate safely, we must be** disciplined and stay within the game plan.

Discussion Questions

1. List some of the dangers that rescuers encounter when responding to a potential suicide jumper.

2. Why must rescuer safety always be the Incident Commander's top priority?

3. List some equipment and procedures that could be used to increase safety at this type of incident.

The Unsung Hero

Battalion Chief Steve
Chikerotis
27 Years in the Fire Service

On this icy winter day in 1994, our company, Squad 2, was rocking and rolling. At 8:00 p.m., we already had worked three fires, and nighttime usually gets even busier. I was the lieutenant of Rescue Squad 2, and we had an outstanding group of firefighters who worked hard, trained hard, and played hard. We liked to go to work, lived for the busy days, and never complained about being up all night.

Our firehouse was never short on laughs, and today was no exception. We had a guest that day; a volunteer firefighter from Iowa named Glen was riding with us. All day long, if we weren't fighting fires, we were busting each other's chops, including Iowa Glen's. My crew really liked this kid, so he was taking his lumps along with everyone else, but I could see that he was enjoying himself.

A Tough Fire

We were sitting in the kitchen analyzing Iowa Glen's love life when the run came in. "You're saved by the bell, Iowa," I said as we ran for the rig. We were dispatched to a still and box (second alarm) in the 9th Battalion. The ground was covered with about a half-foot of snow, but the streets were freshly plowed and the traffic wasn't bad. My driver, Sean, was safe and conscientious, but also adept at getting to wherever we were going.

Listening to the radio traffic, we could tell that this was going to be a tough fire. In just over five minutes, we arrived on the scene. From down the block we could see bright yellow flames dancing out of several windows and lighting up the night sky. As we approached, I studied the building. It was a huge "3 + 1"—a high three-story apartment building including basement apartments. It was about 125 feet wide and 125 feet deep with multiple apartments on the four floors. The building sat on a corner, and fire was blowing out of several windows on the second and third floors.

As we pulled to a stop, I instructed the squad, "Looks like primary search on the third floor. Stay with your team."

Today we had five firefighters instead of our usual six, so we split into two teams. I took Sean with me, and Tommy, Vic and Wayne formed the second team. We grabbed our gear and headed for the building. I approached the Chief, who looked overwhelmed at this point, and told him, "Chief, Squad 2 will search the third unless you need something else."

"Very good, Squad. Primary search on the third floor," he said, and we were on our way.

We followed two hoselines up a stairway that appeared to lead directly above the main fire area. I sent our second team up the next stairway. The smoke was getting thick, and we had to mask up on the first-floor landing. This building had a lot of fire, and I had a feeling we

This is how the fire building looked upon our arrival. (Photo by Mike Prendergast)

were going to find some trapped victims here. We hustled up the stairs, stopping for a minute on the second floor to access the fire conditions.

The engine companies had led a hoseline into each of the two apartments on that landing. I followed the one to the right, and Sean took the apartment to the left. The engine company in my apartment had its hands full with heavy fire in the rear bedroom and kitchen area. I backed out to the landing and met up with Sean as we had prearranged.

"How did it look?" I asked.

"They still have a lot of fire in there," he said.

I told him of my findings, and we continued up to the third floor. There were no lines up to the third floor, so I called the Chief on the radio and asked for an engine company on the third floor.

When we reached that floor, we found both apartment doors still locked—not a good sign. "We might have some people in here," I said to my partner. I forced open the door to the right, and Sean did the same to the door on the left. Heavy black smoke filled both apartments, but the one I forced looked a little worse.

Decision Time

We shut both doors and looked at each other. It was decision time. My mind was whirling. Should we wait for a hoseline? Should we search the apartment to the left? Or should we search the apartment to the right, which seemed to have the worst conditions? With no time to waste, we made a split-second decision—to go with our gut instincts and search the apartment to the right.

> *With no time to waste, we made a split-second decision.*

Opening the door slowly, we eased ourselves into the apartment. We could hear the noise of firefighters working all over the building, but in here it was quiet. The heat was getting more intense as we blindly moved into the darkness. We followed the walls to the right, searching for victims, as we worked our way through the apartment. When we reached the rear, the smoke near the ceiling was starting to light up. This is called rollover, and it signaled us that the whole apartment would flashover soon. We would need a hoseline fast.

I could hear an engine company working outside the kitchen door. I opened it to see the firefighters across the way in a neighboring apartment.

"Stop!" The firefighter on the nozzle yelled to me. He pointed downward, and I saw that the porches between our two apartments had burned away. I was looking down at a 40-foot drop.

"We don't have a line in here. Hit it over our heads," I shouted, pointing upward.

Soon a fire stream was cooling the ceiling gases; we hoped this would delay flashover for a while, and we continued our search. For us, the conditions at the floor were getting worse. The steam created by the hoseline may have delayed flashover, but it also drove heat and dark smoke to the floor.

The next room we came to was just off the kitchen. Sean stayed in the doorway, and I ducked inside to sweep the room. In this type of search, called a "man-at-the-door search," one partner stays at the door and keeps his bearings while the other sweeps the room. At the next room the partners alternate duties.

> *The smoke was so thick that I couldn't see my hand in front of my face.*

The smoke was so thick that I couldn't see my hand in front of my face. Flames flickered across the ceiling as I quickly swept the room with my hands while I crawled around the room.

Emergency!

Then I felt it! Even to a gloved hand, a human body has an unmistakable feeling to it. People don't feel like furniture or pillows. When we find a body, we know it.

I yelled, "Sean, I got one!"

He was at my side in seconds. When I went to grab the victim, I felt another one on top, and then another on top of that one. I soon would realize that we had a pile of bodies stacked about three feet high.

Grabbing my radio, I called in, "Emergency, Squad 2 to 222 [the Incident Commander], we have a whole pile of bodies on the third floor. Get us help!" I followed with our location, then grabbed the body on the top of the pile.

Our second team heard our call for help and tore off a door to bridge the burned-out porch. They were there before I made it out of the

The author (right) and Firefighter Tom McDonough attempted to revive the hero. (Photo by Mike Prendergast)

apartment with my victim. Every firefighter on my company grabbed a victim and followed me down. I had an adult male, a hero, who had shielded his whole family with his own body. Below him were his mother-in-law, his wife, and his four kids. I grabbed my victim under his arms from behind. His back was badly burned from being the human shield, and flaps of skin hung down as I carried him down the stairs.

Finally outside, I set the man down on the snow-covered front walkway. Exhausted and burning up, I ditched my SCBA and firecoat and dropped to my knees to start CPR. I would find out later that immediately after the last victim was removed, the room erupted in flashover.

Triage

I pulled out a pocket mask with a one-way valve and started respirations while another firefighter, Tom, joined and started compressions. The paramedic in charge of the first ambulance had the responsibility for triage at multiple-victim responses, and she knew what she was doing. She quickly assessed that my victim had the least chance of survival, so she went on to assess the others.

A firefighter came over and brought us an inhalator and Tom took over respirations while I switched to compressions. Out of the corner of my eye, I could see Sean carrying a child. He's one of the toughest firefighters I know, and also one of the most compassionate. His face conveyed his pain. Next came Vic carrying a child, then Wayne with another child, and finally Tommy, cradling the mother in his arms. Each of these tough firefighters had the same pained, caring look on his face. Mingled with the emotions triggered by this horrific sight was the overwhelming pride I felt in this group of firefighters.

Other firefighters soon emerged with the other child and the mother-in-law. Each victim to emerge would be swooped up by waiting medics or rushed to an ambulance. Finally, after what seemed to be an eternity, a medic crew came and whisked our victim away to a waiting ambulance. By that time, I barely had the strength to pick up my equipment and walk back to the Squad.

On the way back to quarters, the mood was decidedly different from the mood when we had answered the call. There was no more laughter, no more wishing for another fire. We were physically exhausted and, worse, emotionally drained. Every loss of life is hard to handle, and we knew we lost at least one tonight. Despite our efforts, we knew the heroic father didn't make it, and we were worried about losing some of the others, too.

Left to right: Sean, Vic, and the author critiqued the fire. The grim faces told the story.
(Photo by Mike Prendergast)

The crew returned from the fire. Their pain is obvious (left to right: Tom Greco, Wayne Varney, Steve Chikerotis, Sean O'Driscoll, and Vic Walchuk). (Photo by Glen Pauley)

Upon our return to quarters Iowa Glen asked to take a picture of our crew. We obliged him and tried to pose, but the photo he took conveyed the pain we were feeling.

Talking It Out

Once our air bottles were changed, and our equipment was checked, we went to the kitchen to talk. This was our way of helping each other by letting out our feelings. At the time, we called it a "fire critique" but it actually was a crude Critical Incident Stress Debriefing session. In talking it out, we realized that we couldn't have done anything different to make things better. We had done our best.

While we were still talking, I got a call from an ambulance supervisor, informing me that all of the children and their mother probably would make it. The man's death was confirmed, and there was lingering uncertainty about the mother-in-law, but the rest of the news was uplifting.

That night Squad 2 made a difference in the world. This helped ease the pain, although none of us was looking forward to the next fire. I didn't sleep a wink that night. I stared at the ceiling, thinking. How would this family make it through life without their father? These kids will go through life never knowing the names of the firefighters and medics who saved their lives. I hope they realize how heroic their father was . . . he was truly the unsung hero that night. God bless that family, and God bless my firefighters.

Lessons Learned

- **The priority for searches begins where civilians are most endangered** and proceeds as follows:

  1. fire area

  2. fire floor

  3. floor above

  4. top floor

  5. all remaining floors

- **We should trust our instincts.** Usually, when we go with our gut instincts, we are correct. We form instincts as a result of all of our previous experiences and education coming together.

- **We must search as a team.** The man-at-the-door search is just one type of team search. All searches require practice to succeed. To work as a team, we have to drill as a team.

- **Rollover is one of the warning signs of flashover.** When the ceiling gases light up (rollover), flashover is not far behind. To prevent flashover, the ceiling gases have to be cooled and ventilation added.

- **Rapid and intense heat buildup is also a warning sign of flashover.** If we must belly-crawl and are not on a hoseline, we crawl out, not in.

Discussion Questions

1. Man-at-the-door search is one type of zero-visibility search. Can you name others?

2 Would a formal Critical Incident Stress Debriefing session have been more beneficial to the firefighters in this story? If so, why?

3. What are some factors in deciding where to start a search?

Living on the Edge

Battalion Chief Thomas
Magliano
29 Years in the Fire Service

In our line of work, we never know what to expect. Every day when we report for duty, we are living on the edge. If I've learned one thing in my career, it is to always be ready for a curveball. We don't want to get caught flat-footed. I've had my share of close calls, and I've learned too many lessons to count. The incident related in this story happened when I was a firefighter on Truck 49 in 1986, but the lesson we learned that day is still relevant today.

A Routine Call?

It was late afternoon on a beautiful fall day, and some of us were sitting on a bench outside the firehouse exchanging tall tales. Our shift had been uneventful to this point, but the quiet of the day was shattered by bells ringing and firefighters scrambling for their gear. I was the driver that day.

A few blocks from the firehouse, we saw the smoke. "We got a hit" (a working fire), our officer yelled. Following the engine company toward the building, we started to read the conditions. Heavy black smoke was rolling out of the front door of a clothing store, and a half dozen people were jumping around out in front, pointing us toward the fire.

The clothing store shared a one-story building with two other stores. It appeared to be ordinary construction (masonry and wood joist). The clothing store was on the corner and the other two stores were to the north of it. Each store was separated by a masonry firewall.

As we pulled to a stop, our officer turned to me and instructed my partner and me to "vent that roof."

I stopped the rig directly in front of the building, set the brakes, and put it into PTO (power take-off) for operating the aerial ladder. My assigned partner that day, Denny, was already behind the rig setting the stabilizing jacks as I jumped out of the driver's seat. By the time I was dressed, he had the jacks set. We jumped up onto the turntable, and I raised the aerial.

As soon as we had swung the ladder into position, Denny slung the saw over his shoulder and started up. I followed him with my axe and pole in hand. I raised the ladder to the roof of the center store to keep it away from fire involvement and in a safe location for us.

Denny and I cautiously stepped over the 3-foot-high firewall that separated the two stores. Unless that wall was breached somewhere down below it would act as a good fire stop, keeping the center store from burning. You never know what to expect when you step over to the roof above the fire. With a stiff arm, I slammed the blunt end of the axe straight down into the deck, feeling and sounding for stability. I continued to tap the deck, and it felt solid as we crossed the roof. It looked like the building had a new roof. The seams of the roofing material were about 3 feet apart and coated with a reflective aluminum.

> *I continued to tap the deck, and it felt solid as we crossed the roof.*

Denny fired up the saw with two pulls of the cord. He started his first cut, choosing a spot in the center of the roof about two-thirds back from the front.

Surprise in Store

In reading the building before climbing to the roof, we had noticed that the seat of the fire appeared to be deep in the store. Therefore, Denny's location should be right over the heart of the fire. I headed over to a big exhaust vent about 25 feet away and started to cut at the base with my axe.

As I worked on removing the sheet-metal vent, my back was to Denny and I heard the roar of his saw. The smoke was getting thicker. Denny had penetrated the roof deck with the saw blade and heavy black smoke was puffing out of the cuts. Suddenly flames were everywhere. The whole roof lit up at once, and I was surrounded by fire! Where's Denny? I turned

> *All I could see was a dark figure dancing in a sea of flames.*

toward him, and all I could see was a dark figure dancing in a sea of flames. I had to reach him. I started in his direction, but flames obscured my vision. The fire was now chest high and lapping at my face. I couldn't see Denny.

I was burning up . . . I had to get out of there fast . . . I screamed out Denny's name . . . no answer. I turned for the firewall and started to run. The footing was slippery now, and the smell of burnt rubber was strong. Flames were everywhere; all I could see through my squinted eyes was fire. It was hard to maintain footing on the slippery inferno. I knew that one slip could cost me my life! I was starting to fall as I finally reached the firewall. I dived across the 3-foot high wall and rolled onto the neighboring roof deck.

As I rolled on the deck slapping at myself to make sure I wasn't on fire, I heard a welcome voice. "What was *that?*" came Denny's voice. I turned, and there he was, lying next to me. He was obviously rattled, but like me, he received only a few minor burns. We looked over, and the entire roof was covered in flames that were now about 8 feet high. We made it off that roof just in time. Knowing where the firewall was had saved our lives.

Mystery Solved

We called for a hoseline on the roof, and it was quickly extinguished. The roofing material, not the roof deck, was what was burning. Neither of us had ever experienced anything like that.

Later, we heard about similar roof fires around the city. We wondered what was causing these unusual incidents all of a sudden. We found out that the culprit was a new type of roofing material called "one-ply roofing," "modified bitumen," or "rubber roofs." These roofs, new at the time, had rapidly become the most widely used roofing material for flat-roof buildings.

Research revealed two common hazards with one-ply roofing. When heated, they release a gas that is both flammable and toxic. They have a tendency to flash and ignite during roof operations, because of the flammable vapors collecting near the surface of the roof deck as the roof is heated from below. These toxic and flammable vapors are heavier than air and will accumulate over the deck on calm days when little or no wind is present. They then ignite from open flame, such as when the deck is cut for ventilation.

Lessons Learned

- **We must be aware of the dangers of one-ply roofing and not let one of these roof flashes catch us by surprise.** We always must have a second way off and know where we are at all times.

- **Sometimes we can recognize one-ply roofs.** Because there are several manufacturers of this material, it has different appearances. Commonly, though, the seams are 3 feet apart or wider, usually black or brown in color, with a rubbery appearance. Quite often they are coated with a white or aluminum finish.

- **One-ply roofs aren't always easy to identify.** Therefore we should take extra care when working on roofs.

- **Read the roof upon gaining access.** Always have an alternate way off of the roof. Know the location of the fire and firewalls, and beware of light or airshafts.

- **If you can't see your feet while standing, don't walk.** Crawl, duck walk, or feel in front of yourself with a tool. Never walk blindly.

Discussion Questions

1. What are some construction characteristics of roofs that are of concern to firefighters?

2. What hazards might we encounter while ventilating one-ply roofs?

3. What procedures can we use to increase our safety on the roof?

Not Always Pretty

Lieutenant Thomas
Vogenthaler
15 Years in the Fire Service

What a perfect September evening—clear skies and 70-degree temperatures. We already had about a half dozen runs and a fire under our belt when the call came in: "Squad 1, take in the jumper. Man threatening to jump from roof."

A Race Against the Clock

As the dispatcher gave the westside address, Clarence, our driver, already had turned on the lights and siren and was headed in that direction.

"Vogy and Doti, get a harness on," Jose, our lieutenant yelled to my partner and me. In the crew cab we stripped out of our fire coats, and we each grabbed a vertical rescue harness and carefully started strapping it on.

"Artie, bring both lifelines," the lieutenant barked again. "You and Clarence tie the anchor knots."

"I hope I can remember those knots," Artie joked. It was our way of dealing with the stress of our job.

"Just as long as you get mine right," I retorted. We trusted each other with our lives, so it's no big deal going over the edge of a building knowing that our had partner tied the anchor knot.

On the Scene

"We're about four blocks away. Kill the siren, Clarence," the lieutenant called to us. He didn't want to startle the jumper and cause him to jump.

Within seconds we were silently coming down the alley behind the building.

Suddenly we heard an assortment of excited messages over the radio: "He jumped!" "In the front yard!" "Medics—he's down—let's go!"

We stopped quickly behind the four-story apartment building and ran to the front. We were too late. There he was, impaled on a 4-foot, black, iron-spiked fence. When he jumped, the man had almost cleared the fence, but close doesn't count. Facing the street, his feet had cleared the fence but three spikes impaled his back and were sticking out of his shoulders while his head slumped forward. Implausibly, he was still alive.

With no time to spare, our officer had us grab our Arcair® torch. This torch reaches temperatures of 6700 degrees at the tip but is easy to control and burns through steel like butter.

The ambulance crew packaged the victim, and we cut the fence into a section about as wide as his shoulders. An engine company hit the fence with a fog line as we cut, to dissipate the heat. The torch is an excellent tool for this task because it doesn't vibrate and can cut under water fog, and so the victim doesn't get burned by heat transfer.

On to the ER

In a few minutes, the section of fence that impaled the victim was free. Several firefighters were needed to carefully place the victim face down on the stretcher. We followed the ambulance to the hospital to assist in carrying him in, and also in case they would need us to cut the fence in the emergency room. On occasion, the ER has called on firefighters to scrub and assist in the surgical removal of impaled objects. By chance, an emergency room nurse was riding along with us this day. A friend of my family, she wanted to get a feel for what happens to patients before they get to the ER. She was certainly getting to see a lot this day! This job is not always pretty.

At the hospital we surrounded the stretcher and slowly lowered it to the ground, then quickly wheeled it in to a waiting trauma team. We transferred the victim to the emergency room table, where a medical team surrounded him. We stood back and waited to see if our assistance was needed.

We surrounded the stretcher and slowly lowered it to the ground.

After looking at the impaled back, the doctor asked us, "Fire, can you carefully roll him over onto his back for us?" Slowly, our whole crew lifted both the victim and the fence in unison. Then carefully we turned him over while supporting his body and the fence. As he rolled onto his back, the top of his skull flapped open, exposing his brain.

On our arrival we wrapped the victim's head and other body parts to the fence with gauze to package him for our rescue attempt. Once the fence was cut away, we carried him face up and loaded him in the ambulance. There was very little room for the medics to work back there, so they also were unable to check the back of his head. This caught all of us by surprise.

To our amazement, the doctor leaped onto the table, grabbed the metal piece, yanked it out of his back, dropped it onto the floor, and announced "We won't be needing you firefighters anymore." The priorities for this patient were no longer his impalement. We would find out later that he had survived only for a few hours. I guess he succeeded in his suicide. Not for lack of effort by our crew.

We walked back to the squad, trying to absorb everything we had just experienced. We didn't get that chance. No sooner had we hopped into the rig than the radio crackled again: "Squad 1, are you available?" We were off to another one, a fire this time. Our ride-along nurse just looked on in amazement.

Lessons Learned

● **Do not get tunnel vision.** When working at any type of incident, do not focus on one area. We must see the whole picture. This is equally important when working on a victim.

Discussion Questions

1. What types of safety precautions should be taken on the scene of a suicide jumper?

2. What are some other methods of removing an impaled victim?

3. What are some safety considerations for rescuers working on an impaled victim?

Danger in Disguise

Chief of Department
Greg Brown
24 Years in the Fire Service

Ⓞne of the most important lessons firefighters have to learn is never to lower our guard. In the Fire Service, danger is always hiding just around the corner, waiting for a chance to humble us. Close calls can and do happen. Whether we are new to the job or seasoned veterans, this job tests us. We must learn to expect the unexpected and be prepared to react accordingly.

A couple of incidents come to mind. One of them happened years ago when I was a young firefighter, and another happened to me recently in my current position as Chief of Department.

Catching a Fire

It was a beautiful spring day, the kind you wish you could capture in a bottle. On my drive into work that morning, the sun was shining and the birds were singing. Only a firefighter could look out at all of God's beauty and think, "I hope we catch a fire today." At the time, I had been on the job about four years. This qualified me as one of the most dangerous animals in the jungle. It is said that the two- to five-year firefighter knows everything and can tell everyone else how to do it. I was already comfortable with the job and confident in my abilities.

Shortly after performing our morning equipment checkout, we were dispatched to a reported house fire in a rural area approximately five miles from our station, and we were due first. Within minutes we were approaching an area that you would not call affluent. The houses ranged from 50 to 100 years old, and most were small, with multiple additions.

Each stood on approximately an acre of land. This area had no hydrants, so our water had to be trucked in.

We pulled up to find a two-story frame house with heavy smoke showing. On closer inspection, we found fire venting out of a first-floor window in the rear. The second due truck was still about five minutes away. We had to act fast before the house would become totally involved.

> *The heavy black smoke and heat drove us low onto our knees.*

My captain instructed me to grab the 1¾-inch pre-connected hoseline. Once the line was charged (we had only a 500-gallon tank), he opened the front door and together we attacked the interior. Immediately, the heavy black smoke and heat drove us low onto our knees.

At this point, everything seemed to be business as usual. As we worked our way to the rear, it got tougher. The residents had the habits of packrats, and we had to maneuver through narrow aisles of boxes and stuff piled all the way to the ceiling. The house was like a big maze, which increased the danger. The clutter would add tons of fuel to the fire and slow our progress.

A Humbling Experience

When we were about halfway through the house, conditions worsened. It was getting hotter by the second, and the black smoke was pumping faster. We could see the glow of fire in the rear. The second company had arrived, and my officer radioed that we were having difficulty reaching the fire.

Firefighters from the second company stretched a second 1¾-inch hoseline to the rear. The fire coming out of the rear window was now igniting the eaves and spreading to the roof, so they decided to attack the fire that was showing in the rear. When they opened up a fog stream on the fire, the stream covered the rear window, which cut off our natural ventilation. The intense heat built up quickly.

We didn't know what hit us as the conditions rapidly became unlivable. We were flat on the floor and still felt like we were melting. Quickly we did a U-turn in the narrow aisle and, sliding along in a belly-crawl, we tried to find our way out. The pile of stuff surrounding us was making escape difficult. Like black fire, the smoke was incredibly hot. We were stinging from the heat and knew we had to get out fast.

In an instant, I was no longer a cocky know-it-all. I was a scared young firefighter fighting for survival. Boxes had collapsed across our hoseline, making the line difficult to follow outside. My ears and the back of my neck stung like a thousand bee stings. This happened in the days before we wore hoods.

> *I was a scared young firefighter fighting for survival.*

A Second Chance

Following an outside wall, my captain found a window, and we quickly broke it out. Firefighters from the second company were about 25 feet away, working their line when they saw us at the window. They shut down their line and ran to help us. With their line shut down, the fire again began auto-venting out of the rear window, and within a minute the conditions started to improve again. The natural ventilation was working wonders, and once again we turned to attack the fire.

> *The natural ventilation was working wonders.*

Now we were able to push the heat out the back window, and the smoke began to lift off of the floor. They inadvertently had been pushing the heat down on us in two ways. They had blocked our ventilation point and created great amounts of steam with their fog line. By driving the smoke down to the floor they also created zero visibility.

With the improved conditions it was a different ball game. We started to make good progress. Within a few minutes we were able to find the door to the bedroom that was on fire. A pile of boxes partially blocked it, which made it hard to locate with zero visibility. Crawling low, we made it through the door and knocked down the fire. Other firefighters came in to overhaul the ceiling and walls. Our job was finally completed.

Feeling the effects of the burns, we exited the building as fast as we could. Our helmets were damaged, our face shields had melted completely, and even the reflective trim on our gear had melted. Our burns turned out to be minor first- and second-degree, and we knew how lucky we were to have escaped. The incident made me look at fires differently, and I became more of a student of the job.

No Time to Duck

The other incident happened recently. Now the Chief of the Department, I was at my parents' home in my staff car when I heard an engine company being dispatched to my sister's house for a car fire. It was close, and of course I responded.

The first to arrive, I saw that the engine compartment on my nephew's Jeep Cherokee was fully involved with fire and a large amount of fuel was burning on the driveway. The vehicle had been pulled up close to the garage and the fire was spreading to the eaves. The back hatch of the Jeep was open because someone had been retrieving items from the rear of the vehicle.

I pulled a 10-pound dry chemical extinguisher out of the staff car and discharged it on the garage and fuel. When the engine company arrived, the fire had spread to the passenger compartment. The crew pulled a pre-connected hoseline and extinguished the eaves, then turned their attention to the Jeep and started to extinguish the flames that were rolling across the ceiling of the passenger compart-

> *I heard a loud pop, like a gunshot.*

ment. I was walking behind the Jeep when I heard a loud pop, like a gunshot. Then something whizzed by my head so close that I felt the breeze as it flew by. It hit the officer side door of the fire engine.

Rethinking Vehicle Fires

Upon further examination, there was a 1-inch-diameter hole in the center of that door. Inside the cab of the engine, we found a gas strut that had held the back hatch up. With the back hatch open, the natural draw pulled the fire to the back, and what appeared to be a minimal amount of fire caused the strut to BLEVE (boiling liquid expanding vapor explosion) and turn into a projectile.

The strut had just missed my head and, with the force of a shotgun blast, went completely through the engine door. I was about 20 feet behind the Jeep, and the engine was about 25 feet from me.

This was a close call that I never saw coming. It made me rethink how we fight vehicle fires. Since then, I have heard of gas struts blowing through a closed car hood in an engine compartment fire. We now train our firefighters to attack vehicle fires from a side angle, as the struts tend to shoot directly out of the front or back. On this job danger sometimes is disguised.

Lessons Learned

- **We must never block a ventilation hole.** The first incident shows that playing fire streams into ventilation holes (windows or roof openings) creates steam and a reverse airflow, which pushes heat back into the building.

- **A secondary means of egress is essential.** We have to know where we are in relation to the exit and also have a secondary means of exit.

- **We have to approach from an angle and cool from a distance.** The vehicle incident showed that gas struts exposed to fire are most likely to fly directly in front of and behind a vehicle, so we should approach at an angle from the side. In approaching, we should cool the vehicle with the fire stream from a distance.

Discussion Questions

1. What is your Fire Department's tactical procedure for structural fires similar to the house fire in this story?

2. What are some modern tools and equipment that would add to the protection of firefighters working at a similar structural fire?

3. What is your Fire Department's standard operating procedure for vehicle fires?

Losing My Way

One thing about firefighters: We're human, and we all make mistakes. Mistakes common to the fireground include freelancing, communications breakdown, failure to read the building, having a false sense of security, and tunnel vision. Every now and then we receive a much-needed wake-up call that reminds us of our vulnerabilities. Preferably, it doesn't end with an ambulance ride. I received one such wake-up call on a cold Sunday morning.

Firefighter Quentin Curtis
17 Years in the Fire Service

A Bad Omen

The alarm came in shortly after our 8:00 a.m. roll call. Dispatch was sending us to a working fire in a multiple-unit apartment building. I had a gut feeling about this one. When we catch a fire early on a Sunday morning, it's usually a bad omen. "It's Sunday morning. This is the real deal," I remarked as we hopped aboard the squad. On Sunday morning calls, we've had a lot of fatalities. Maybe it's because a lot of people do stupid things on Saturday night. Also, most people are at home and most are sleeping late. This tends to put a high number of people in a vulnerable position.

Squad 1 made good time through the light traffic and arrived in a few minutes. A large volume of heavy black smoke shrouded the building as we pulled up. This made it harder to read the building, a large U-shaped, courtyard apartment building. It was abandoned, and all of the doors and windows were boarded up tightly with plywood. Heavy smoke was puffing through cracks around the plywood on the third-floor windows, which indicated that we had a large fire on that floor.

Even though the building wasn't formally occupied, this neighborhood had a lot of homeless people who sometimes took up residence in these abandoned buildings. Two hoselines were stretched into the center doorway, another line went into a second door, and the first truck had raised its aerial ladder to the roof. We grabbed our tools and went to work.

Our lieutenant quickly divided our crew of six, sending two firefighters to the roof to ventilate, and my partner and me to the third floor to perform a primary search. The lieutenant and his partner would search another section of the third floor. Each team had a portable radio, which my partner carried. The two of us, dancing over the charged hoseline, hustled up the center stairway.

> *The two of us, dancing over the charged hoseline, hustled up the center stairway.*

The smoke was thick as we reached the second-floor landing. We kept going until we reached the top floor. Although the black smoke completely obscured our vision, we could hear several firefighters working and followed one of the lines into an apartment.

Everywhere we turned, we seemed to bump into a firefighter. There are way too many of us in this apartment, I thought. We should check the second floor. I tried to find my partner, but he had disappeared into the blackness. My voice, muffled by my face piece, couldn't compete with the sounds of saws, axes, and fire streams. I took off by myself.

Freelancing

About a week earlier, we had arrived at a second-floor fire in a two-story house and joined the first engine and truck in attacking the fire and searching the second floor. This resulted in a delay searching the first floor, where heavy smoke had banked down. A dead child eventually was found in a first-floor bedroom, and I blamed myself for the delay. Tunnel vision had taken us to an area where everyone else was already working.

With this incident fresh in my mind, I headed for the second floor. When I reached the landing, the smoke was thick and the visibility was zero. I forced open the door of the apartment that was directly below the main fire area. Heavy smoke was banked to the floor. Heat, though not intense, also was present, indicating a small fire somewhere on the floor.

The apartment was cluttered with furniture and boxes. When I checked one of the bedrooms, I encountered a mattress and blankets on the floor. This often indicated that homeless people had taken shelter. I scooted through the apartment in a duck walk, keeping in mind that syringes are commonly found on the floor in abandoned buildings.

Now I was deep in the apartment and the smoke was getting hotter. I was tripping on clutter and getting turned around until it hit me: I didn't know where I was. I'd better get out.

Window to Safety

For several minutes I tried in vain to find the stairway. As the heat increased even more, I caught myself starting to panic, and I stopped in my tracks. Calm down and don't panic, I told myself. Thinking clearly again, I started following the walls and tapping with my fist.

Finally I felt something hollow. Plywood. I've found a window! I chopped at the plywood with my axe and finally was able to pry it off. Tapping with my axe, I felt glass, and using the butt of my axe as a ram, pushed through the glass and at the same time popped the plywood off the outside of the window.

Sticking my head out of the window, I frantically waved my arms to draw attention. Heavy smoke encircled my body and now was rolling out of the window over my head. Once I saw firefighters point in my direction, I knew a ladder would be there soon. Heat now was being drawn to the vented window, so I stuck my handlight on the windowsill and ducked below the opening to avoid the heat.

I frantically waved my arms to draw attention.

Within a minute I heard the sound of steel on brick as the tips of the ladder crashed against the windowsill. I gave my brothers a few seconds to adjust the ladder and then I popped out of the window and onto the ladder. I couldn't have lasted much longer inside there. My air supply was about to run out, and the smoke would be lighting up at any second.

I was surprised to find out that I was the last firefighter out of the building. A large section of the roof had collapsed, and the Chief had ordered everyone out of the building to start a defensive attack. I was unaware of this because my partner had the radio. In the confusion and zero visibility, Joe hadn't even noticed that I was missing.

This was a much-needed wake-up call for me. After that fire, I became much more safety-conscious. I had received a firsthand lesson on free-lancing, communications, accountability, and tunnel vision. The thing that saved me was that I had taken time to read the building, and I knew that the windows were covered with plywood. Also, I resisted the urge to panic, stayed calm, and thought my way out of there.

Lessons Learned

- **Freelancing is never acceptable. Stay with your partner at all times.** Also, firefighters should never change locations without informing the officer.

- **Communication is the key.** Verbal communication with your partner is important in zero visibility. Radio communications, firefighter to officer and officer to Chief are essential for accountability.

- **Roll call must be taken after a building evacuation or a Mayday call.** All officers must be polled via the radio after an evacuation or Mayday: "Do you have all of your people?"

- **Defensive tactics must wait until everyone has been evacuated.** We can't mix defensive and offensive tactics. Defensive tactics utilize indirect firefighting, which requires a large amount of steam conversion, and this can be deadly to firefighters inside. Also, the use of master streams increases the chances of building collapse.

Discussion Questions

1. What is your Fire Department's tactical procedure for evacuating a building?

2. What is your Fire Departments tactical procedure for a Mayday call?

3. What are some survival techniques that Quentin could have used?

Play Your Gut Feeling

Chief Robert McKee
32 Years in the Fire Service

During my thirty-two years in the Fire Service, I've been blessed to work alongside some great firefighters. My father was a Chief, and my brothers and I followed in his footsteps. An excellent role model, our father instilled in us the importance of being a student of the job and sharing our knowledge with others.

Now a Battalion Chief, I still take his advice to heart and will remain a student until the day I retire. I've also been an instructor and teacher throughout my career, knowing that the information we pass on to others may save their lives. The incident I relate here happened on a cold November day in 2001.

Reading the Conditions

At 3:00 a.m., with a foot of snow on the ground, I maneuvered my way through the empty streets. I saw flashing lights converging on the upcoming intersection. It was a truck company headed to the same still and box (second alarm fire), so I slowed and let the truck turn in front of me. I followed behind, trying to stay in its tire tracks. There wasn't much radio traffic, so I had little information about the fire. Because of the heavy snow, it took almost ten minutes to get there. When I arrived, I buried the car in a 2-foot drift of snow and jumped out to grab my gear.

As I walked toward the building, I began reading the conditions. This was a two-story, ordinary construction (masonry and wood joist), "taxpayer" structure—apartments over storefronts. It was about 150 feet wide and 100 feet deep. Dark black and brown smoke was pumping out of three of the four storefronts.

The engine companies had already led out three lines to the front. Two lines disappeared into the storefronts, and one stretched up a front stairway to the apartments. The aerial ladder pierced the thick smoke at roof level, which seemed to swallow the ladder. I could hear the saws of a busy truck company venting the roof. The Deputy District Chief was the Incident Commander, and I found him standing in front of the

> *The aerial ladder pierced the thick smoke at roof level.*

building. "What do you need, boss?" I asked.

"Go around back, Bobby, and let me know how it looks in Sector Three" (rear of the building).

I trudged through the deep snow to reach the rear, reading the conditions on my way. Around the back of the building, Engine 122 and Tower Ladder 34 had stretched a 1¾-inch line up an enclosed back porch to a second-floor apartment. I entered the back-porch area, where the smoke was thick all the way to ground level. When I masked up and started up the stairs, I could hear the bells of firefighters SCBAs going off. If they were low on air, it meant they'd been up there for close to fifteen minutes already, and conditions were getting worse. I could still hear the saws buzzing above us, but the smoke didn't seem to be lifting. Something wasn't right.

Something's Not Right

I found the captain of the tower ladder, and he informed me that they had found only small pockets of fire.

"Let's get some of this ceiling down," I said, "and make sure it's not over our heads."

He ordered one of his firefighters to pull the ceiling in the kitchen where we were. With a couple tugs of his pike pole, a large section of ceiling dropped. Within seconds, the conditions went from good to bad, and then worse. The ceiling was full of fire, and it was blowing straight down at us.

"Let's go! We're backing out of here!" I yelled.

As the last firefighter exited, the flames were down to the floor. Heavy black and dark brown smoke was rolling into the enclosed back porch. The firefighters worked together in an orderly retreat. When we reached the bottom of the stairs, I instructed the tower ladder company to force the rear door to a furniture store, which still had heavy fire and smoke conditions.

The firefighter on the nozzle shot the stream up at the ceiling, and the water just disappeared. Nothing happened to the fire. We aren't making a dent in this fire, I realized, and backed our group away from the building.

I called the Incident Commander: "Battalion 22 to 226 [Deputy District Chief], we're backing out of the building in Sector Three. The structure is burning too badly on the first and second floors."

The Deputy District Chief ordered everyone out of the building. We were going to set up defensive operations. About five seconds after we cleared the rear of the building, we heard a loud "Ka-Boom!" Two-thirds of the second floor had collapsed into the first floor.

> *Most of them had to drop their tools and run for their lives.*

In the front of the building, firefighters dived out of the doorway and narrowly escaped. It was so close that most of them had to drop their tools and run for their lives. Four members of Squad 5 were cut off from the stairway and were trapped on the second floor. Hearing their call for help we directed a truck company to quickly raise a 28-foot ladder to their window. An engine company covered their escape from the street by driving back the flames that danced over their heads. One by one, they popped out of the window, until all four had escaped.

The Deputy District Chief initiated an accountability roll call, and each company officer announced that all of their firefighters were accounted for. This one was way too close. Thankfully, every firefighter made it out of the building safely.

The Chief called for a 2-11 (third alarm), and we fought it defensively for the next four hours. Several secondary collapses happened throughout the firefight, but all firefighters were safely out of the collapse zones. In talking to the companies after the fire, I had a clear message: "Tonight we got away with one, let's make sure that we learn from it."

Afterthoughts

In reflecting on this fire, there were definitely lessons learned and lessons reinforced. I messed up in the beginning because I didn't get enough information from the Incident Commander before I headed to the rear. I should have asked him how long the fire had been burning and what actions had been taken already. If I had I done this, I would have had a better handle on the conditions.

A lesson that definitely was reinforced was to trust our gut instincts. Another lesson reinforced was to pay attention to time: How long has the fire been burning? Also, we always have to open up the building, know where the fire is, and find out where it's going.

Last—my dad was right: Always be a student of the game.

Lessons Learned

- **When gut instincts surface, we must trust them. They're usually right.** The instincts that Bobby McKee displayed that night were honed from crawling hallways and learning from every incident. Instincts also are developed by attending classes, reading books, and from firefighters sharing experiences. The knowledge that we gain is filed away in our toolbox that we call a brain. It's always there for us. Don't forget to use it!

- **Some conditions that can trigger instincts are:**

  - **Time:** The building isn't getting stronger as it burns. From the sound of bells, the Chief in this story realized they had been there at least fifteen minutes and weren't making progress.

  - **Sound:** The sound of sawing without the smoke lifting told him that something was wrong. False ceilings were preventing the heat and smoke from venting.

  - **Smoke color:** The dark brown and black smoke indicated that this wasn't just a "contents fire." The structure itself was burning, and ordinary construction (masonry and wood joist) has collapse potential in less than twenty minutes after structural involvement.

  - **Volume:** The larger the volume of fire, the longer it has been burning. The longer it has been burning, the closer it is to building collapse.

Discussion Questions

1. What are some other indicators of potential collapse?

2. Why is time a consideration at structural fires?

3. Can you name other characteristics of smoke that can affect our tactical decisions?

Teamwork

Captain Rick Kolomay
26 Years in the Fire Service

On this job, our greatest challenges do not always come from fires. One of the most challenging tasks that firefighters perform is extrication. The incident in this story was an industrial extrication that required skill, training, and teamwork if we were to succeed.

Industrial Extrication

In October of 1999, at 4:45 a.m., we were called to the stock area of a supermarket. Upon arrival, I performed a size up and found a nineteen-year-old male in the sitting position on the floor. His left leg was crushed at the thigh by a cardboard compactor. The crushed leg dangled into a pit, pinched under thousands of pounds of pressure by an 8-foot-square sheet of diamond plate steel. He had slipped while loading cardboard into the compactor. Before his leg was completely severed, a quick-thinking employee jammed a broomstick into the machinery, which tripped a breaker and stopped the machine. The victim's thigh was now compressed to about an inch thick.

Going to Work

We arrived with a task force consisting of a Chief, one engine company, one truck company, one ambulance, and my squad company. Our immediate task was to stabilize both the victim and the machine quickly. The medics quickly put the victim on oxygen and started a line. At the same time, we deenergized the compactor and wedged two haligan bars to stop any further movement. The truck company opened the sides of the

> *The Chief orchestrated the whole rescue effort.*

compactor and unloaded the cardboard. The engine company dropped a 1¾-inch line to cool the steel and protect the victim if we had to end up cutting with a torch. The Chief orchestrated the whole rescue effort.

The first attempt we made at extrication was to lift the diamond plate with the Hurst Tool or Jaws of Life®. The tips kept slipping out, and they wouldn't budge the huge steel plate. We then tried to use a 4-ton and a 7-ton airbag to lift it. At the same time, the truck used the chain saw to remove wood slats inside the compactor. This would expose the diamond plate steel so we could cut it with our Arcair® torch if the airbags failed. At the same time, I instructed Mark, the best mechanic on our squad, to climb up on top of the machine to see if he could figure out the mechanics of it.

What Next?

The two airbags failed to budge the diamond plate. It was starting to look like we would have to cut the steel with the Arcair® torch. The torch reaches 6700 degrees at the tip, so the fog line would have to be used in conjunction with it to dissipate the heat. Mark had determined that the compactor was gear-driven and not hydraulic. He asked me to pass him a 30-inch haligan bar. The Chief, who was supervising all operations from

> *The plate started to lift slowly.*

inside the large compactor, was ordering the use of the Arcair® when we saw movement in the plate. The airbag that was wedged into the small opening fell into the pit. The plate started to lift slowly.

The victim screamed out in joy when the pressure was relieved. Mark had been able to slowly turn a large gear that was lifting the plate off the victim's leg. Shortly, the victim was taken to the hospital, where he was determined to be in stable condition.

A Job Well Done

The entire operation took about 1½ hours, but everyone was so busy that it seemed like ten minutes. We had taken every safety precaution for both the victim and the rescue workers. As a team, we simultaneously used all of our resources to extricate the victim as quickly and safely as possible. The Chief orchestrated one of the most professional operations I had ever witnessed.

In this incident no new extrication techniques were utilized and no heroic efforts were needed. The formula for success was a well-coordinated team effort.

Lessons Learned

● **Seeing the whole picture.** This extrication was successful because the highly trained crews saw the whole picture—the opposite of tunnel vision.

● **Each accomplished an assigned role,** with an emphasis on victim and firefighter safety.

● **Professionalism comes from training.** Having the most sophisticated tools and equipment means nothing if the rescuers aren't proficient with them. We need to train to be proficient.

Discussion Questions

1. What is the first step in performing an industrial extrication?

2. What other type of challenges can firefighters face in industrial extrication?

3. List some safety measures that can be taken to protect the rescuers?

Becoming a Student of the Game

Over a thirty-year career, we experience many wins and losses. This story is about a day that changed my career. I've chosen to tell this story in the hope that it may help the next brother or sister put in a similar situation. At the time, I had been a firefighter on Squad 5, a heavy rescue squad, for several years.

Lieutenant Pat Lynch
30 Years in the Fire
Service

Just Another Day

The day was like any other day. By 10:30 that night we had just returned from our third fire of the day and our crew was sitting around the kitchen table having coffee. The conversation was really about nothing, and we were listening to the scanner with one ear.

Our conversation was interrupted when we heard companies reporting on the scene of a still alarm. Engine 92, Truck 45, and Battalion 21 gave descriptions and reported that nothing was showing. We resumed our discussion and were in the process of solving the world's problems, when we heard an urgent voice: "Battalion 21 to Englewood emergency give me a box" (second alarm).

"That's us—let's go," the guys said, running to the rig. Not much was being said on the radio, but by the tone of the Battalion Chief's voice, we could tell something wasn't right.

It took us about seven minutes to get to the scene of the fire. When we were within a few blocks, we could see a thick column of black smoke and the glow of a fire. As we arrived, one of the chiefs came running up to us, screaming, "Get in there and get them! Firefighters are missing!" I'll never forget the look on the Chief's face.

I'll never forget the look on the Chief's face.

The building was already lost at this point. It was a tire store with a one-story showroom area tied into a shop area about 75 feet by 150 feet with a bowstring truss roof. A single 2½-inch hoseline led into the front door.

We didn't know how many firefighters were missing—somewhere between two and four, we were told. An accountability check was still under way. Time wasn't on our side. The entire structure was full of heavy flames. A large overhead door to the shop area appeared to be breathing fire. The lower 6 feet of the doorway was sucking in air in a rhythmic fashion, and the top half was belching flame and black smoke to the same rhythm. "This looks like the gates to hell," I thought. With these conditions, coupled with the truss construction, the roof could collapse at any time. We had to act fast.

Before We Arrived

When the first engine and truck had reported on the scene, they found a tire store; no fire was showing through the large showroom windows. The business was closed for the night and locked up tight as a drum. The firefighters walked around to the back of the building and looked into the shop area windows for signs of fire and saw none. Shining their hand lights into the darkened service area, nothing seemed out of the ordinary. A neighbor who lived across the alley approached and told the firefighters that he saw the flicker of flames in the rear shop window and called 911. As the group was starting to force entry to investigate, the owner came running up with keys. He had been notified by a private alarm service.

As they entered through the showroom door, the crews of the first engine and truck were assaulted by the smell of burning plastic or rubber. One of the officers relayed a message to the Chief: "It smells like we have a car fire in the back."

The crews passed through a set of double doors into the shop area, where the smell was stronger, but they still didn't see any fire. Beams from hand lights scanned the shop. Half of a dozen cars were back there, two up on hydraulic lifts. The crews spread out to find the source of the smell. High above the floor in the bowstring truss area, the smoke was starting to get thick, but at floor level it was totally clear.

Where was the smoke coming from?

Where was the smoke coming from? One of these cars has to be burning, seemed to be the general consensus. They continued to search for the source. Unknown to the group, above the showroom a fire was raging in a hidden storage room. The smoke and heat were pumping into the bowstring truss area of the shop. The temperature at the roof level had to be above 1300 degrees, and the barrel-shaped roof was quickly filling up with smoke, carbon monoxide, and other fire gases. Several feet below, the firefighters had a false sense of security because they were still able to walk around, felt no heat, and had good visibility. This was about to change.

Flashover!

Within a few minutes the firefighters started to lose their visibility as the smoke banked down toward the floor. An overhead door was opened and suddenly, "Flash!" The whole ceiling lit up orange. The onslaught of heat drove the firefighters down to the floor. The dreaded flashover had occurred, and firefighters scrambled to escape this blast furnace.

Panic was his first instinct.

One of the men was a young firefighter who had just graduated from the Fire Academy. This was his first fire, and panic was his first instinct. He started to run blindly through the superheated smoke.

The voices of his Academy instructors sounded in his head: "Stop! Firefighters don't panic. Hit the deck, stay low, and go." These words of advice had been drilled into his head in the Academy. He listened.

When he hit the deck, the conditions on the floor were much more livable. Now under the thermal layer, he could see nothing but darkness in the direction he had been heading. Even at floor level, he felt like he was going to melt.

This is the showroom window through which the firefighters escaped the flashover.

"Firefighters don't panic—they react," continued the voices from his memory. He spun his body around on the floor until he saw the flicker of lights from the fire truck through the showroom window in the distance. Staying low, he scurried in that direction. Just before he reached the window, another firefighter dived through it, and the young rookie followed closely behind. Flames like a blowtorch followed them out the window without a second to spare.

> *Flames like a blowtorch followed them out the window without a second to spare.*

Our Squad on the Scene

The quickly deteriorating conditions prompted the Battalion Chief to call for a second alarm. This is the point at which our squad company was starting out for the fire. After the flashover, and as firefighters dived to safety, the Chief quickly took a roll call via the radio. It became apparent that some of the firefighters were unaccounted for. When we arrived, the rescue effort was underway. Firefighters were scrambling to bring extra hoselines into position.

Three of us from Squad 5 followed the 1¾-inch hoseline in search of the missing firefighters. Hearts pumping, forcible entry tools in hand, and with the best intentions in the world, we headed in. About 40 feet into the building, we found the nozzle end. No one was on it, just a charged line to the doorway of the shop area. Smoke and heat conditions were getting worse by the second.

We had a 75-foot by 150-foot area to search in zero visibility with strong collapse potential. These are insurmountable odds, but the missing ones were our brothers. One firefighter stayed at the door working the hoseline at the trusses, which seemed as futile as a garden hose flowing into a lake. My partner and I sank eye-level to the ground, trying to see anything as we headed in. We were listening for a PASS alarm or any sign of life, but all we could hear was the crackle of fire.

> *All we could hear was the crackle of fire.*

We crawled between two rows of cars, praying that we would hear a cry for help or stumble across a body, but we had a huge area to cover. "We should have brought the rope bag," I said, and we called for it on the radio. My partner Jimmy and I backed out to the hoseline and waited for one of our squad members to bring in the rope.

Against the Odds

Three or four minutes went by before were able to tie off and reenter. One firefighter stayed on the hoseline, and back we went, to the same area. This duplication of effort wasted more time. Having no formal proactive game plan for a wide area search, we fumbled around, searching, lifting up our masks, and yelling. Still no reply.

We knew the roof was going collapse. The question was: How long do we have before we die? Do we have minutes or seconds left? The vibra-alert on our face piece regulator was telling us that we were almost out of air. Under normal conditions, an experienced firefighter could get twenty minutes out of an air bottle. Under the conditions here, it would be less. Running out of air was probably a good thing. My instinct was telling me to

> *My instinct was telling me to get out, but my heart was telling me to stay.*

After the collapse, Squad 5 deployed a master stream from the snorkel.

get out, but my heart was telling me to stay. Out of air, we had to retreat. Boy, do I hate that word! Retreat. But sometimes we have to retreat or more firefighters will die.

Once outside, we ran to the squad and changed our air bottles. When we got back to the door, the Chief stopped us: "That's it—nobody is going back inside. This building is going to collapse." We knew that would happen, but by this time we also knew that two men were missing. "We can't just give up!" Other crews were breaching the masonry walls and trying anything they could to locate their brothers.

A few minutes later, "Crash!" The trusses collapsed with a thunderous roar. Flames shot into the night sky. At that moment I felt I had failed—and failed big. And I had thought my skills and knowl-

> *The trusses collapsed with a thunderous roar.*

edge were among the best. Even after many years and many fires, a firefighter can get a curveball and fail. It hurts.

The recovery operation lasted for several hours, hampered by the constant barrage of secondary collapses. Two brave firefighters died that night. One of them was found deep in the shop area. Firefighters had to breach a masonry wall to recover his body. One of these heroes was later recovered no more than ten feet from our last position, buried under

After the collapse, firefighters breached the masonry walls in an attempt to find the missing firefighters.

debris. We had just missed finding him when the search had to be called off. Was he already dead at that time or could we have still saved him? Was there more that we could have done? These questions will always haunt me.

Since that day, I attend every class I can and read every book, article, and report I can get my hands on. I became an Instructor for the Fire Service Institute. Most important of all, I stay humble. The best way to teach is not to blow my own horn but to share my failures. In my teaching role, I have traveled around the country and met some great people. We in the Fire Service share a common bond: to protect life and property. I take great pride in being a student of the game, and it will never end. God bless.

Lessons Learned

- **A piece of good advice is always to be a student of the game.** In this story, a seasoned veteran of a busy company found that he had more to learn.

- **High ceilings can cause a false sense of security.** High ceilings, such as bowstring trusses, can give a false read on the severity of the fire and heat conditions. We must not be fooled.

- **The Fire Service is continuously changing.** In the few years since the incident reported in this story, several Departments across the country have added Rapid Intervention Teams (RITs) and RIT training. Integrated PASS Alarms now are built into SCBA units. Thermal Imaging Cameras (TICs) are becoming standard.

Discussion Questions

1. Why was the initial attack crew unaware of the high heat conditions?

2. What are some of the warning signs of flashover?

3. At a similar incident, what modern tools or equipment could be used to prevent a similar tragedy?

Some Valuable Lessons

Deputy District Chief
José Santiago
27 Years in the Fire Service

Over the last twenty-seven years, I've seen a lot of action and learned many valuable lessons. My partners and I have made mistakes. Sometimes we paid dearly with painful injuries, and on other occasions we were lucky and escaped close calls. Over the years I've learned from my mistakes and realized the value of teaching others not to make the same mistakes.

The Mentor

Our Fire Department had an age-old tradition of letting the children of firefighters visit the firehouses. Over the years some of the lucky ones were able to ride along and observe at fires occasionally. One such child was John, the son of one of our chiefs in the early years of my career, when I was a firefighter on Rescue Squad 2.

> *Unknowingly, we became his heroes.*

Young John immediately developed a liking for the job, and his father used to bring him around quite a bit. One might say that he grew up in the firehouse. He used to ride along with us, and over the years we watched him grow up. John's dream was always to be a firefighter "just like you guys." Unknowingly, we became his heroes.

Rescue Squad 2 was a busy company and at the time full of young, overly aggressive firefighters. Our experienced captain used every trick in the book to harness our energy productively. I know we were a handful for that poor captain. We did work hard, but at times our never-back-down attitude made us careless. Sometimes we escaped injury as a result of our captain's wise supervision and leadership, and sometimes we just lucked out.

The New Recruit

Years passed, and John achieved his dream. When he entered the Fire Department, I was one of his instructors at the Fire Academy. After graduating from the Academy, he was assigned to Truck 35, a busy company in our area. I was promoted to lieutenant and assigned as a relief officer in the same area. Most days, John and I worked at fires together.

One summer evening at about 6:00, we responded to a fire in a three-story ordinary construction (masonry and wood joist) apartment building. I was the officer of Engine 43 that day. By the time we arrived, heavy fire and dark brown smoke could be seen coming from the back of the building. Circling to the rear of the building, I found the outside rear porches were fully involved with fire from the ground all the way to the roof level canopy. This roof stopped the vertical spread of the fire from rising into the sky. The orange flames were curling and lapping right into the window and door of the third floor kitchen area.

"Bring a line around to the back, we'll attack it from the rear," I called to my company. We stretched our hoseline to the rear of the building, knocked down the fire on the porches, and started working our way up to the third floor.

About halfway up I stopped; something didn't feel right. My instincts told me that this porch was about to collapse. The fire had burned for too long and weakened the porch to an unsafe condition. We were now adding too much weight

> *My instincts told me that this porch was about to collapse.*

and impact load. "I don't like this," I yelled to my company and the truck company that was starting to follow us.

"We'll go back around and attack it from the front," I added. We started back down the stairs when I heard someone yell, "What are you guys doing? Keep going. We'll get it." It was the familiar voice of young John, who by now had a few years on the job.

As the crew backed down the stairs, John worked his way up to me. "What are you doing? We have this licked."

I said, "No, it's too dangerous and not worth the risk."

He looked at me and said, "No way. When you were on Squad 2, you never would have backed off." As a child, John developed a lot of confidence in me. Now he had the look of a kid who just found out there is no Santa Claus. He stared at me. "What happened to you? You never backed down in the old days."

I barely paused, then made a split-second decision. "You're right. Let's go!" I told the others to stay down because there would be too much weight for the weakened porch to bear any more people. John and I grabbed the line and started back up to the third floor. Just as we reached the top, it felt like someone pulled the rug out from under us. The porch had given way, and the two of us had an express ride to the basement.

> *The two of us had an express ride to the basement.*

A Wake-up Call

I couldn't believe it was happening to us. As if in slow motion I remember trying to grab onto anything to stop the fall. I was so relieved when the ride finally stopped and there was John beside me. Within seconds firefighters were clearing debris off of us and dragging us out of the pile. "We need an ambulance," I heard several of them scream.

"Is John okay?" "What have I done?" I used up my get out of jail free card that day. John and I were lucky that we escaped serious injuries, but this incident was my wake-up call. It was time for me to grow up and become a real leader.

As I sat in the hospital, I reflected on this incident. "What if this kid had gotten killed?" I had violated every leadership principal I had learned in the Fire Service and in the Marine Corps. As a firefighter, I took risks that involved only me, but as a company officer, I was charged with the responsibility for the health, welfare, and guidance of an entire company. No longer could I think of what people would think or say. Bravado has no place in the life of the seasoned officer I thought I was. A company officer must never let ego come in the way of safety, sound judgment, and professionalism.

Lessons Learned

- **If we have a bad feeling, we're probably right.** We must err on the side of safety.

- **We should avoid fire-weakened stairways.** Because collapse usually is triggered by impact load, on any stairway we must spread out our weight (one firefighter per section of stairs), spread out our feet (step directly over the stringers), and avoid impact load (walk, don't bounce).

- **Ego has no place on the fireground.** Instead, we must base our decisions on risk management. Egos have gotten many firefighters killed.

Discussion Questions

1. What other tactics could have been used in attacking the fire in this story?

2. What guidelines should be used in determining if the attack should be offensive or defensive?

3. What conditions probably triggered the collapse?

He's Got a Gun!

O ne thing I learned on this job is to always expect the unexpected. I firmly believe that the lessons we learn throughout our careers can save our lives in the future. In the first incident I relate here, I had been on the job only about three weeks, which paved the way for my response to a similar incident fifteen years later.

Rescue Supervisor
Howard Nolan
20 Years in the Fire Service

Foreboding

On that steamy afternoon the cloudless sky seemed almost too blue to be real. Our engine company was rolling down the street, returning from a medical run. The officer and driver were in the front, and I stood in the open crew cab, sun beating down on me, wind in my face. The roar of our old diesel engine disrupted the quiet of the day as we rolled down the street. Heads turned and children waved. What a job!

Heads turned and children waved. What a job!

The beauty of the day was about to change, as the real world reared its ugly head. It was in the form of a dirty yellow Oldsmobile and its occupant. The car pulled alongside of us. We were in the left lane, and the car appeared to be trying to pass us in the right lane—except it didn't pass.

It became apparent that the car was attempting to keep pace with us, which struck me as unusual. I looked down at the driver. His left hand was on the steering wheel, and in his right hand he held what looked like a gun with a 12-inch barrel. I watched the surreal image as he took aim at us. "He's got a gun!" I yelled.

Bam! Bam! Bam! The stream of gunshots and the old diesel engine drowned out my screams. I hit the deck and watched the muzzle flash and smoke from the large barrel of what appeared to be an Uzi. The car swerved violently as the bullets flew over our heads. After several shots had whistled by in a matter of seconds, the driver accelerated and passed us at a high rate of speed.

I bent over and stuck my head close to the officer's window. "The guy in that yellow Olds just shot at us!"

"He what?" the officer shouted back in disbelief. He and the driver had no idea what had happened.

Positive Identification

When the officer called our dispatch, we were about a block or so behind the yellow car. It made a left turn that would take it right past our fire station, and then past the police station, which was about five blocks down the road. An off-duty firefighter who happened to be in the fire station ran outside when he heard our message, and there was the car, sitting behind traffic at a stoplight. The firefighter ran back in and informed dispatch, and within seconds four police cars had cornered the offender.

We pulled over by the firehouse and waited to hear from dispatch. A few minutes later we were told to meet the police at that location.

There, an officer approached and asked me, "Did you see the gun?"

I was able to give an accurate description of the Uzi automatic machine gun.

He asked, "Do you want to see it?" and took me over to the car. There was the gun, lying on the passenger floor. I was asked to identify the shooter, and we headed back to our station. The incident was the talk of the town for a while: "Did you hear?" "A brand new firefighter got shot at by a machine gun!"

From this experience, I learned to expect the unexpected. I also developed the philosophy of "safety before all else" and to never lower my guard because danger can come from anywhere. This incident made me more safety-conscious from that day forward. It also prepared me for a similar incident that happened years after this one.

Danger can come from anywhere.

Fast Forward

Fifteen years later, I was the Rescue Supervisor on Rescue 4. One evening we responded to a medical call for a man having a seizure. We pulled up in front of a two-story townhouse to find a woman screaming frantically, "He's upstairs! He's upstairs!"

Grabbing our medical equipment, we started to follow her to the house. Before entering, a neighbor came up and the woman stopped to talk, so we passed by her and headed up the stairs. I was the third in line behind my two firefighters. A few steps from the top landing, a man appeared above, aiming a large-caliber handgun at us.

From my perspective, the barrel looked like a cannon aimed between my eyes. The man's eyes had a crazed look and appeared to be bulging out of his head as he screamed, "What are you guys doing in my house?"

> *The man's eyes had a crazed look.*

His hand shaking in anger, he again questioned our presence. We were in uniform, carrying a med-kit and oxygen bottle. Calmly and quietly I told my two guys, who were still ahead of me: "Hold it. We're backing out. No sudden moves. Walk backward one step at a time."

The man continued to aim the gun at our heads and yell at us as we slowly backed out. I could see that the hammer was cocked and his finger was twitching on the trigger. By this time his wife was yelling at him, but he was still enraged and seemed not to hear her. We continued to back away slowly until we were out the door.

Then I radioed for the police, and we got back in the squad and headed down the block until our back-up arrived. At that point, the wife led the police into the house with their guns drawn, and they disarmed the man of a loaded .357 Magnum.

I have no doubt that if we had acted aggressively in any way or had frozen in place, the man would have shot one or more of us. We quietly left before he even realized what was going on.

Experience—The Best Teacher

Without a doubt, the incident fifteen years prior helped me react as I did. I probably was able to stay calm and think logically because it wasn't the first time I had to stare down a gun barrel. Also, the first incident made me more safety-conscious in everything we do, and it turned me into more of a thinking firefighter. Now, my emphasis every workday is on the safety of my firefighters.

> *Sometimes safety is learned the hard way.*

Sometimes safety is learned the hard way. I was lucky not to get shot in that first incident twenty years ago, but I became a better firefighter and officer. I also became a Fire Academy Instructor, and we preach safety, safety, and more safety.

When we train new firefighters, we go overboard in teaching them to *think*. At fires, we want them to slow down and spend more time at the front door before they go in. We want them to ask themselves, "What could be happening on the other side of the door?" By taking those extra couple of seconds to think, they're more likely to act in a safer manner than by going in recklessly.

More than 100 firefighters die each year in the United States. The case studies reveal that between one-third and one-half of the deaths are directly related to safety issues. One life lost needlessly is too many.

Lessons Learned

- **Civilians panic, and firefighters react.** This adage gives us job security. A cool head helps get us out of some tough situations.

- **We should never take anything for granted.** We can never lower our guard. This job is too dangerous to relax.

- **We must think safety.** The author of this story said it best: We must always think, act, preach, and teach safety. Thinking firefighters save lives—sometimes their own.

Discussion Questions

1. Why is training and drilling so important in the fire service?

2. What are some ways in which fire departments can be proactive about safety?

3. How might a firefighter make a difference?

The Pentagon on 9/11

Chief Robert Klinoff
30 Years in the Fire Service
Author of *Introduction to Fire Protection*, Third Edition/Thomson Delmar Learning

September 11, 2001—a day that would change the world—started just like any other day. This would quickly change as the terrible events began to unfold. Word spread around the globe like wildfire, and soon the eyes of the whole world were watching in horror. Our Interagency Incident Management Team was no exception. As I saw the unthinkable scene unfold on live television, I waited for my phone to ring. Soon it did. A dispatcher called to notify me that our group was being activated.

I keep a duffel bag packed at the ready for just such calls, and within a few hours we were in the air. At that time, we were aboard the only private plane in U.S. airspace. We knew we were headed to the East Coast, but we didn't know our final destination. We were only told that it would be either the Pentagon or the World Trade Center.

Dispatched to the Pentagon

We made one stop in Albuquerque, New Mexico, to pick up another Incident Management Team. While in flight, our team was informed that we would be assigned to the Pentagon. The other team would be going to the World Trade Center. We were being escorted by two fighter planes to the Baltimore–Washington International airport.

As the plane made its landing approach, I looked out the window, and off in the distance I could see a thick column of smoke rising from the Pentagon. On the ground we were met by armed U.S. Marshals and

The interagency operational base camp can be seen at the bottom of the picture. Rescue efforts were still under way at the time the photo was taken.

loaded onto buses. We were staged in an area for a short while and then transferred to the base of operations at the Pentagon. We set up our operations in a grassy area about 300 yards from where the plane had hit.

Our team was enlisted to support operations with planning and logistics and to help the urban search-and-rescue teams in any way we could. I was assigned as a safety officer. We were working in conjunction with the various branches of the military, the FBI, Secret Service, FEMA and USAR teams, the EPA, and the Arlington County Fire Department.

Our Role

Our first major assignment was to bring a sense of organization to the operation. We were the last of the agencies to arrive, and the tents had been set up in no specific order, which made coordination among the various agencies difficult at best. We helped set up a unified command center with each of the agencies represented to coordinate the effort.

Second, we supported safety both inside and outside of the Pentagon. This is probably the most secure military site in the world, and we had to pass through five levels of security to enter the work area. Armed military personnel monitored our every move. The threat of another attack on the same site was constant because it is such a strong symbol of the military strength of the United States.

Hundreds of rescue workers aided by rescue dogs tunneled and dug through the debris in search of victims. The site was noisy with heavy equipment and power tools being operated in several areas, but every once in a while there would be total silence in stark contrast. Every time a victim or a body part was located, all operations were halted. The FBI then would document the area where the body was found before we were allowed to remove it. At those times, an eerie silence fell over the entire area. For the first time in my life, I truly understood the meaning of "the silence is deafening."

> *An eerie silence fell over the entire area.*

We realized that few, if any, survivors would be found, despite our efforts. We never gave up hope, but as the hours and days went by, we really knew what we didn't want to believe. It was apparent in the grim yet determined faces of all of the workers. The eyes of everybody involved in the operations conveyed sadness.

The hardest part for me was in knowing that, as firefighters, we usually are in a position with the potential to save lives. That's what firefighters do, but at this incident it was not to be. The hopelessness of the situation wore us down emotionally.

After twelve days we were returned home. As we loaded up our gear, we walked away with an empty feeling that we didn't fulfill our purpose.

Hard Lessons

This incident taught us a lot of lessons. One is that we have to see the whole picture as a large-scale incident is developing. We must recognize its potential and react accordingly. Because the scene is changing constantly, our size-up must be continuous. We have to know where this incident is going and plan ahead. If our Incident Management Team had arrived earlier instead of being staged after others were already on the scene, we could have used our organizational expertise in setting up a location for each of the other agencies. Instead, we were always playing catch-up in trying to organize.

> *Our team motto was "quiet confidence."*

Another lesson we learned is to recognize the strength that each agency brings to the table and never let ego drive our decisions. Our team motto was "quiet confidence."

We can make changes by showing others what we can do to make things better. This has to be done humbly, not by walking up and saying "We're here to take charge." By utilizing each other's strengths, we are all working toward a common goal and accomplishing it more safely and efficiently. Terrorism creates a crime scene with a HAZMAT and it requires several agencies working together.

A lesson that was reinforced is that in a large-scale incident such as the one at the Pentagon, the World Trade Center, or a forest fire, the conditions are out of our control. In cases such as these, we can work for several days and go home without any resolution. We just have to do our best and be able to accept the result. Most firefighters are of the mindset that they are going to make things better, that they are going to resolve the issues, and then they will leave. Unfortunately, it doesn't always work like that.

Dealing with Emotions

Coping with the emotional stress that comes from dealing with so much death and destruction requires follow-up. Before we returned home, we had Critical Incident Stress Debriefing (CISD) sessions with the military. This helped us greatly in handling our emotions. Also, it was heartwarming to see the nationwide patriotism. We returned to neighborhoods, where every house was flying a flag.

The whole nation was rallying together and being supportive, and this helped the whole nation heal. Our team also had a welcome-back party from our family members, and this was much appreciated. We will always recognize the importance of firefighters and our families looking out for each other.

Lessons Learned

- **A priority for Incident Management Teams, in particular, is to see the whole picture.** When we arrive at the scene, our size-up tells us the current stage of the incident. The next question is: "Where is it going?" We must recognize the potential and plan accordingly.

- **We have to check our egos at the door.** At all incidents, we must be willing to utilize the strengths of other agencies to accomplish our goals.

- **We must look out for ourselves and others.** Because of the nature of our job, we always must look out for the emotional well-being of each other and utilize Critical Incident Stress Debriefings when needed.

Discussion Questions

1. Why would it be beneficial to have one Incident Management Team coordinate the positioning of all agencies at large-scale events?

2. At large-scale incidents, what are the benefits of having a unified command center with all agencies present?

3. What are some of the benefits of Critical Incident Stress Debriefings?

G. I. Jane

Joining the Fire Department was a dream of mine for as long as I remember. This wasn't something I wanted to *do*. It was something that I wanted to *be*. My philosophy is that if we want something in life, we don't just sit back and wish for it. We go out and work for it. My motto is: "Do nothing in moderation."

I have always been an athlete, and athletes have a lot in common with the Fire Department. If I have learned one thing through sports, it is that the harder we work, the greater is our success. I trained harder for the fire test than I ever had for an athletic competition. All of my hard work paid off.

Firefighter Meg Ahlheim
5 Years in the Fire Service

Starting a New Career

On July 17, 2000, I walked into the Fire Academy to start my new career. The adrenalin rush was like the one I used to get on game day, only more intense. Our class of 125 candidates (new recruits) was divided into twelve groups. The group proved to be a family within the family. This prepared me for when we left the Academy and my fire company is a family within the bigger family, the Fire Department.

> *The group proved to be a family within the family.*

From the first day, I could tell that the Fire Academy was going to be tough and challenging. I also sensed more camaraderie than ever before. We recognized that in this job we have to cover for each other in life-and-death situations.

The Birth of G.I. Jane

On the morning of my first day of Academy, we were inspected and dressed down by a Chief who was also a Marine. He chewed us out for our appearance. Some of the guys in my group had long hair, sideburns, or a mustache. At the end of our first day, the eleven recruits in our group had our own meeting and decided to do something about it: "Let's all go get our heads shaved in a buzz cut."

When I came through the gate the next day, I must have shocked a lot of firefighters. One pointed at me and shouted, "Look—it's G.I. Jane!" That name stuck.

Fortunately, everyone else in my group also came in having shaved their heads. Our group made a statement that day, and we formed a true bonds.

It didn't matter to me that I was only one of seven females in the class. I didn't want to be good female firefighter. I wanted to be a *good firefighter* period. On the second day we had a PFT (Physical Fitness Test). I ended up number eight overall, and the highest female. This wasn't good enough for me. I wanted to be the best.

Because I had done so well, some of the guys targeted me. They weren't going to be beaten by a girl again! Well, I like a challenge, and it made me work all the harder. I trained above and beyond the physical demands of the Fire Academy, coming early and staying late every day. My motivation was to be the best I could be. I befriended a building engineer so I could sneak into the building to work out on weekends. My husband came with me and timed me with a stopwatch as I ran the six-story fire escapes.

> *My motivation was to be the best I could be.*

During our six months in the Fire Academy, I grew as a person and our class grew as a family. Our close-knit group studied together, worked out together, and watched out for each other. We pulled for each other and truly wanted each other to succeed.

Each month we had another PFT, and each time I crept closer to the top. On the final PFT, I ended up number two in the class, and I received an award at our graduation ceremony. My family beamed with pride.

On-the-Job Demands

After graduating I was assigned to Engine 5. My first week on the job, we got called to a fire in a high-rise, and we had to walk up the stairs to the twenty-eighth floor carrying equipment. Two workdays later we were called to a big magnesium fire, where we shoveled sand and salt for about five hours. Adrenalin can take a person only so far. Physical conditioning is essential to be effective on this job.

Also, it became apparent that, as part of our conditioning, cardiovascular training is necessary because of the nature of our work. Professional athletes wake up at a certain time on game day, eat a prescribed pre-game meal, stretch and warm up before the competition, and hit the field prepared. By contrast, firefighters jump out of bed at 2:00 a.m., hop on the rig, and two minutes later are running up stairs and crashing in doors. This puts a tremendous strain on the heart.

Coming Back as an Instructor

With a little more than a year on the job, I was asked to come back to the Fire Academy for a six-month detail as an assistant physical fitness instructor. Prior to coming on the Department, I had been a physical trainer for two years, so I was comfortable with my ability to take on this assignment. Still, I felt a little awkward coming back and working alongside people who had been my instructors one year earlier. I needn't have been concerned. The other instructors quickly made me feel like part of the family.

Even though motivating the new candidates was rewarding, I couldn't wait to get back to the Fire Company again because I missed the action. I discovered a way to enjoy the best of both worlds, and currently, in addition to my firefighter duties in the field, I'm part of the Fire Department's Wellness Committee. This committee develops programs that promote good health for all members of the Fire Department. Also, I have been named the Fire Department's representative to the Mayor's Physical Fitness Council, which deals with public health issues.

Pride

In the Fire Service we take pride in everything we do, and having an active, healthy lifestyle plays a big role. I used to train to compete

===

We come to work to do a good job.

===

athletically, and now I train to do a good job. At work, we support and trust each other, and we don't want to let our company down. We don't come to work with an ego where we have to prove to anyone that we can do a good job. We come to work to do a good job.

I have learned that this is a tough occupation, and it isn't for everyone. Along with a dedicated body and mind, it requires commitment of the heart.

Lessons Learned

- **Physical conditioning and eating a proper diet have to become part of our lifestyle.** Firefighting is among the most dangerous occupations in the world, plagued by injuries and deaths. As we improve our health, we are better able to avoid injuries and maybe even save our life.

- **We have to set obtainable goals and work toward them.**

- **We can be a role model for others** by staying fit and promoting a healthy lifestyle. We owe it to ourselves and our families.

Discussion Questions

1. What are your personal goals for your own physical fitness?

2. How can a fire company set goals and promote health as a company?

3. What are some job-related exercises firefighters can do at work as part of a company drill?

We Don't Always Win

Battalion Chief Steve
Chikerotis
27 Years in the Fire Service

The year was 1992, and I was one of the luckiest men on earth. I had it all. I was living a dream life, with two great families—one family at home and every third day my family in the firehouse. I was the lieutenant of a busy heavy rescue squad company, Squad 2, which responded to fires daily with rescue as our primary mission. We also responded to vehicle extrications, hazardous materials incidents, vertical rescues, scuba-diving rescues, building collapses, and any other type of emergency that came up.

Squad 2 ran with one officer and five firefighters. The average firefighter in our company had more than ten years on the job. Each was hand-picked for this squad because of their skills, qualifications, enthusiasm, and heart. With this group of firefighters, it was fun to come to work. We worked hard and we trained hard because we took pride in being the best—and I believe we were. Because we had been together for several years, we had come to know each other like family, and I respected and trusted each of my firefighters. I would go to war with this crew. Actually, we did go to war together, every third day.

Assigning the Duties

On a cold October morning around 3:00 a.m., the city was asleep, but not us. We were rolling down the street doing what we liked most—going to a fire. That might sound strange. We really don't wish that people's houses and businesses would catch fire. It's just that if they do, we want to be there.

Tommy drove through the empty streets while the rest of us prepared our gear. We heard the first engine company arrive, and the officer described a large fire in the first floor of a six-flat apartment building. Considering the time of day and the location of the fire, I turned to Tommy and said, "This could be bad."

Sizing up the conditions as we rolled to a stop, I made the assignments: "Tommy, set up the snorkel in Sector One and then hook up with Sean and Vic on Three. Wayne, help the truck on the roof. Duff, it's you and me on Two."

Our teams were set. We always organized teams at the morning roll call, but there were invariably last-minute adjustments. We had to play the hand dealt to us.

Thick black pressurized smoke was puffing out of all six apartments of the three-story building. We could see the rich orange glow of fire through the thick smoke in the first floor windows. No occupants were in front of the building, and no victims were visible in the windows. Where were they? While quickly walking to the front door, our group studied the building as if our lives depended on it—because it did. We masked up and followed the first engine company's hoseline into the building. As we expected, the visibility was zero, and it was hot.

Primary Search

Inside the first-floor apartment, it sounded like a bull in a china shop as Engine 44 attacked the fire, but the chirping of smoke detectors was conspicuous by its absence. "No detectors—not good," I yelled to nobody in particular. The chance of finding victims had increased tenfold.

As our team headed up the front stairs, I radioed the incoming Battalion Chief with our progress and status. The heat increased with each step. This would be a tough one, I told myself. On cue, I heard breaking glass from above, and glass from the rooftop skylight rained down on our helmets. "Nice job, Wayne," I thanked him silently.

Glass from the rooftop skylight rained down on our helmets.

Instantly the heat on the stairway started to lift and conditions were much more tolerable. As we reached the second-floor landing, Duffy forced the door with his pry bar. We were directly above the fire

Squad 2 (left to right): Wayne Varney, Tom Banks, Vic Walchuk, Sean O'Driscoll, and Steve Chikerotis. (Photo courtesy of Jim Regan)

apartment. As we slid inside to start a primary search, the smoke and heat had us banked to the floor. In a half-crawl, half duck walk, we followed the outside wall deeper into the apartment.

Our necks bowed and our heads withdrawn like a turtle, Duff and I tried to escape the heat. Our ears stung as if we had run into a swarm of bees, but we pushed on. Finding victims quickly would be the only way to save a life here. Any delay would turn this into a body recovery operation. We were here for one purpose: to save lives. I led as we quickly swept the front room and headed down the hallway. My left hand on the wall, I swept in front of me with my free hand, and Duffy was close behind. I heard him sweep the floor, using his axe handle like a blind man's cane.

My left hand felt a doorjamb. We had reached our first target. With no smoke detectors and during normal sleeping hours, my instincts told me we were going to find someone here. "Bedroom!" I yelled back to Duffy.

> *My instincts told me we were going to find someone here.*

I felt behind the door as we slid into the room. We passed a dresser and hit the outside wall. I felt for a window. Bingo—there it was. "Window!" I shouted to Duffy. Within seconds I heard my 220-pound partner making short work of the window frame with a pick head axe. I hit the next wall and felt a hinge. "Closet," I announced, quickly searched it, and moved on. I felt the end of a double bed. Leaning across the center of the bed, I flapped both arms and felt the bed top to bottom. Duffy rejoined me as I was sweeping my leg under the bed. Nothing but shoes. "The room's clear, Duffy."

"Windows vented," his muffled voice replied. We pushed on.

The second bedroom produced the same results, and within five minutes we had finished searching the apartment. We found the rear door ajar. This was how the occupants got out, we guessed.

Exhausted and overheated, we collapsed against the back wall. I slapped my partner on the back. "Nice job, Duff!"

The conditions started to improve slightly. The engine company was winning the battle, and Wayne and the truck company were making Swiss cheese out of the roof. Sean radioed that Vic, Tommy, and he had the same results on the third floor. "Squad 2 to Battalion 6, primary search is complete on Second and Third, Sector Two side, no victims."

Secondary Search

The Chief answered, "Very good. Start a secondary search."

With improved visibility and heat conditions, we searched the entire apartment again, in reverse order. This meant searching each room thoroughly for victims and fire extension. By the time we reached the first bedroom again, we were standing up and had taken off our face pieces. Duffy was rechecking the closet as I pulled back the mattress.

"Thud." It was an unmistakable sound. I threw the mattress and box spring to the side. Oh my God—I missed a child!

"Squad 2 to Battalion 6, we need an ambulance!" I yelled into the radio. "We found a child on Two, and we're bringing him down the Sector One stairway."

My partner and I grabbed the boy, who looked to be about twelve, and ran him down the stairs. When we laid him on the front porch, he wasn't breathing and had no pulse. We started CPR. His face was covered with soot, but he wasn't burned. I cleared his airway and started mouth-to-mouth resuscitation while Duffy did chest compressions. "Breathe, please breathe," I pleaded.

Soon our paramedics took over and rushed the victim to the ambulance. I don't remember much after

"Breathe, please breathe," I pleaded.

that, except that we went to the hospital after the fire was out and got the bad news that he didn't make it.

It Still Hurts

Twelve years have elapsed, and my whole body feels numb as I relive the pain. I blamed myself for not finding the boy during the primary search. After this happened, I was in a depressed state for a long time and didn't talk to anyone about it. I was quiet at home and quiet at work. I hugged my own four sons every chance I got. I saw that little guy's face constantly, and I lived in guilt.

Critical Incident Stress Debriefings weren't common in 1992. At that time, I thought I didn't need any help. I was too tough for that stuff. I was wrong. Over time I came to realize that I had done my best under those tough conditions. It still hurts, though.

The only time I had sat in on a CISD was when a group of us returned from New York City after 9/11. I lost some friends in the Trade Center and like everyone else I was hurting inside. Our Fire Department chaplain set up a debriefing for our team, and I reluctantly attended. I came away from that debriefing with a whole new appreciation for CISD; it helped tremendously. I left feeling like someone had lifted an anvil off my chest.

Lessons Learned

- **If we give it our best, we can live with the results.** We work in the toughest conditions imaginable, and we aren't always going to succeed.

- **Primary search is a rapid search.** Time is the most essential element in rescue. We must search each room quickly if the victim in the next room is going to have a chance.

- **Secondary search is a more thorough search.** Besides, the conditions and visibility are usually improved.

- **CISDs really work.** We should watch for signs of critical incident stress in our co-workers and utilize CISDs whenever needed.

Discussion Questions

1. How thorough should a primary search be? Why?

2. How thorough should a secondary search be? Why?

3. What type of team searches does your Fire Department use?

Trapped in a Burning Stairway

In July of 1986 I followed in my father's footsteps and entered the Fire Department. I idolized my father, Roy Dean, Sr., and being a firefighter was all I ever wanted to be. To make things even better, I was assigned to Tower Ladder 14, a busy company. My career has been a continual learning experience, but one of my biggest wake-up calls came when I had been on the job about three years.

Lieutenant Roy Dean
19 Years in the Fire Service

Fire on a Hot Day

On a 90-degree summer afternoon we were sitting in the kitchen talking. It was a quiet day, and we hadn't had a run for a few hours. Around 3:00 p.m. our laughter was silenced by the voice on the speaker: "Engine 117 and Tower Ladder 14 take in the still alarm" (first alarm). The house bells were ringing, and we followed engine 117 down the street.

As soon as we hit the apparatus floor, the heat of the day descended on us. Our firecoats were still wet from sweat as we slipped them on. The odor assaulted our noses, but the smell of smoke from our gear balanced it out. This is one smell we crave.

We saw the column of black smoke from a few blocks away. When we pulled up, the first engine company was just starting to lead a hose into the front door of the two-flat building. Heavy smoke was rolling out the front door and puffing out of several windows. The first truck company was raising its aerial ladder to the roof when we pulled to a stop on the corner. The lieutenant detailed to our company for the day ordered, "Grab the 38 footer, and let's ladder the rear."

As we approached through an alley at the rear of the building, heavy smoke filled the air. We raised the ladder quickly to give the roof team a secondary means of egress, and we headed into the enclosed back porch. From there, we hustled up the seven steps to the first-floor landing.

> *We could hear pots, pans, and dishes crashing around the room.*

One of our firefighters rammed the rear door with the butt end of his axe, and the door flew open. Smoke and steam started to fill the stairway. We could hear pots, pans, and dishes crashing around the room from the hose stream as the engine company worked inside the kitchen.

Off on My Own

"Seems like they got it knocked. Let's open up these windows," the officer commanded. The fire seemed to be coming under control.

I slipped off by myself and went up to the second-floor landing to open the windows. The second-floor porch had three windows and a door with a closed security gate. I decided to open the windows first and then force entry into the door. I set my axe and pike pole in a corner and started opening windows. My bulky gloves made it impossible to open up the triple-track storm windows, so I pried them off and threw them beside my tools.

I had half-opened the first window when suddenly the smoke turned black and hot. In an instant, conditions turned from good to bad and were getting worse by the second.

> *Flame was rolling up the stairway.*

Then—flash!—everything lit up on fire. Flame was rolling up the stairway. I dived for the base of the door. How was I going to get through the security gate?

Firefighters were yelling my name from the bottom of the stairs, and I didn't even have time to answer. I grabbed the security gate, and it swung open. Luck was with me—it was shut but not locked. I wasn't out of this predicament yet, though. I shouldered open the wood entry door, and inside the second-floor kitchen the heat conditions were much better. I shut the door behind me to buy some rapidly dwindling time and started crawling in zero visibility toward the front door. I bumped into some kitchen chairs and fought them off until I found the wall.

Just as I started to follow the wall, I heard my officer in the front of the apartment shouting: "Roy Dean, where are you?"

I answered, "I'm right here. I'm okay."

He yelled back, "I have the front door. Follow my voice."

As conditions rapidly worsened, I was happy to comply. We were united on the front landing.

Too Close for Comfort

Gratefully, I asked him, "How did you get here so fast?"

He explained that a bedroom window next to the first-floor door broke as the room flashed over. The flames raced up the stairs and drove the guys to the floor. Thinking he had lost me, he did the only thing he could do—ran through the first floor to the front stairs, which were protected by a hoseline. He was about to come in and search for me when I answered his voice.

This one was too close. It gave me a whole new appreciation for the dangers of the job. Sometimes in the first few years as a firefighter, we get too comfortable and take safety

This one was too close.

for granted. I realized the importance of using protected stairways when going above a fire. I also learned to plan for a secondary means of egress at all times. I should have checked the security gate first. If it was locked, I would have been badly burned there—or worse. On that hot summer day, I became a student of the job.

Lessons Learned

- **We should utilize protected stairways to go above the fire.** A protected stairway is the one from which the engine company is attacking. The company can't protect a stairway in the direction of the push.

- **We must always have a Plan B.** We have to be aware of our surroundings and have a secondary exit. How do I get out of here if my original path is blocked?

- **It can never be said too often: Stay with your partner.** Freelancing kills. We must stay within verbal communication of our partner at all times.

Discussion Questions

1. How would a Mayday procedure and Rapid Intervention Companies help in a similar situation?

2. What advantages does a protected stairway offer?

3. How could communications and accountability be improved in a similar incident?

Be Ready When You're Called On

I've been a student of the service ever since I came on the job. I've learned from some outstanding officers and firefighters and attended every class I could get into. The lessons have come early and often in my career. My goal has been to be prepared at all times to meet the many challenges we face on this job. I want to be ready.

One of my first real tests came on a hot, humid August night in 1990. At the time, I had been on the job only about four years and was a firefighter on Rescue Squad 1.

Captain Kevin Krasneck
19 Years in the Fire Service

Reading the Building

On that day, I was the driver, and when I pulled up to the fire building, I was already sizing up the conditions. It was a two-story wood frame building. The 30-foot by 75-foot building housed a tavern on the first floor and a dance club on the second floor. Dark gray smoke was puffing from the street level door to the tavern.

The street was filled with civilians who had gathered to watch the first-to-arrive companies search for the source of the fire. The officer quickly deployed our squad company, and off they went to work. Being the driver, I had to jump into my turnout gear before hooking up with my team on the first floor.

While I quickly dressed, I heard the first engine company report finding the fire in the basement. In the minute it took me to dress, the conditions had worsened considerably. Heavy dark brown smoke was now rolling out of the tavern door, a color that I had learned to recognize as wood burning. This meant that this wasn't just a "contents fire." It was a structure fire. With my axe and pike pole in hand, I headed for the front door.

Just before I entered, the Chief grabbed me. "Open this wall over here. I want to see if the fire is traveling up to the second floor." He took me to a second door in the front of the building, from which a straight-run stairway led to the second-floor dance club. He wanted to have the outside wall opened, so I went up about four stairs and opened the wall with my axe. Heavy dark brown smoke started puffing out and rising up the stairway. I turned to report this information to the Chief when I was almost run over by a firefighter charging down the stairs.

> *Heavy dark brown smoke started puffing out.*

A Missing Partner

I was surprised that anybody was upstairs. I thought it had already been searched and evacuated. This firefighter clearly had run out of air while doing a primary search and was now at the bottom of the stairs gasping for air.

"Is anyone else up there?" I asked.

He nodded. "Yes, I'm missing my partner." I looked up and the stairway was now filled with smoke that was getting thicker by the second.

I reported to the Chief, who was standing in front of the building about 10 feet away from me: "Chief, a truck company is missing a firefighter on Two, and I'm going up to find him."

"Go ahead. I'll get you help," he responded.

Quickly I masked up and headed up the stairs into the darkness. I knew that the straight-run stairway would take me about 20 feet deep on the Sector Four side (right side of building). It was getting hotter as I climbed the stairs and there was no visibility, only total darkness.

From reading the building before entering, I noticed three double-hung windows in the middle of Sector One (front of building). I filed that information in my head, because that could be a secondary means of egress if needed. I had learned long ago that we always have to prepare for the unexpected on this job.

When I reached the top of the stairs, I yelled, "Is anybody up here?" I stopped breathing to silence my exhalations, and I listened closely. I heard a muffled yell and some banging noises in the rear on the Sector Two side (left side of building).

"Stay where you are and keep making noises. I'm coming for you!" I shouted, then started a lefthand search (left hand to wall) and headed back toward the front wall. I knew that it was about 20 feet to the front and that I cover about 3 feet on each crawl. In my calculation, I would reach the front wall in about seven crawls. Once at the front wall I continued my lefthand search and started feeling higher up on the wall for the three front windows. I found only one window. Had I overshot the windows? Was I in another room?

"I'm coming for you!"

Quickly I opened the window and stuck my head out to see where I was. I was in the center of the three windows. The other two windows had been covered with drywall on the inside. I shut the middle window because I didn't want to feed the fire with more oxygen. Then I marked that window by leaving my pike pole on the floor pointing out into the center of the room, and I continued my search. Five crawls, or about 15 feet later, I reached the outside wall of Sector Two. Now I knew exactly where I was as I followed the missing firefighter's sounds deep into the room.

While I followed the wall, I kept bumping into overturned chairs and tables. The truck company might have done this during the primary search, and it was making my own search difficult. I had been taught to move furniture off a wall only if we have to search behind it, and then to push it back against the wall.

I kept bumping into overturned chairs and tables.

Found Him!

Crawling as fast as I could, I reached the lost firefighter. My search to this point had taken less than five minutes. He was flat on the floor and out of air, his face pushed to the carpet.

> *Buddy-breathing off my mask, we made our way to the front.*

I pulled off my mask and gave him some air. Telling him where we had to go, we reversed to a right-hand search (right hand against the wall), which would take us back to the front stairs. Buddy-breathing off my mask, we made our way to the front.

He was starting to show signs of panic, and I told him to relax, that I knew the way out. We hit the front wall and I knew the window was only 15 feet away. This was a good thing because by now we had run out of air. The heat was increasing by the second, driving us tightly to the floor. As we inched our way along, trying to suck air out of the carpet, I kept a hand firmly wrapped around my new partner's shoulder strap.

Soon I felt my pike pole, and I knew we were right below the window. I opened it and we sprang to our knees and popped our heads outside. I could hear my partner wildly sucking in the air. I could feel my own chest rising and falling; the fresh air tasted so good that it took a while for me to realize we weren't safe yet.

"We have to go," I told him. "Just a little farther and we'll reach the stairs." He wouldn't budge. I told him that we were only about 30 feet from the stairway, but after being so close to death he didn't want to give up his window.

I radioed the Chief, told him that I had the missing firefighter, and asked for a ladder at the front window. Then, driven by the heat, we dropped to the floor and we waited.

A minute later I rose up onto my knees to see if help was on the way and a ladder came crashing through the window. It struck me on top of my helmet, jamming it down into the bridge of my nose. I started bleeding, and my nose would require stitches later. The firefighter dived out onto the ladder, and I followed as the smoke lit up behind us. As we descended, I looked up to see flames now blowing out the window. We had made it with no time to spare.

We found out later that a rescue effort had been underway for us. An engine company was protecting the stairway with a hoseline, while a team searched for us. This effort was called off once we were accounted for.

Reinforced Training

This incident reinforced the training that I had received. I saw firsthand the importance of reading a building on arrival—how it helped me navigate blindly through the second floor. It also reinforced the importance of search-and-rescue techniques. These skills are not only for saving others' lives. They also can save our own. I also saw how utilizing landmarks can help us navigate through zero visibility. Finally, this incident reinforced the importance of training and drilling.

The only way we are going to be able to perform well when called upon is to practice, practice, and practice some more. Thanks to the officers and firefighters who broke me in, I was ready when I was called on.

I made some serious mistakes that night also. Due to the seriousness of the situation, I made a split second decision to go in alone; and, I failed to tell my officer where I was. I also should have radioed the Chief immediately after finding the victim and given our location.

We were lucky that night. Other than my little cut, no firefighters were injured at this incident. This occupation is way too dangerous to rely on luck. We must learn from our mistakes. I know that I did.

Lessons Learned

- **Companies should have a Mayday procedure and practice it.** This incident happened before a Mayday procedure was instituted in our Departments. Now, most departments have one, but it will work on game day only if we practice.

- **Accountability and communication go hand in hand.** When a higher-ranking officer tells us to do a different task, we have to let our officer know where we are and what we are doing.

- **Good officers have to be good teachers.** When a four-year firefighter like the one in this story performs like this, you know he's had some good teachers.

Discussion Questions

1. What is your Fire Department's Mayday procedure?

2. How do Rapid Intervention Teams add to the safety of the fireground?

3. What modern-day tactics and equipment could be used in a similar situation if it were to happen tomorrow?

A Deadly Department Store Fire

Matt Woody
President EMS Resource
Group
28 Years in the Fire Service

Not a cloud was in sight. The morning sun illuminated a bright blue sky that served as the backdrop to a beautiful Sunday morning. Leaves of gold, red, and orange floated down from branches of the mostly bare trees, which, despite the mild weather, signaled that summer had come and gone. It was November 9, and the holidays were just around the corner.

I was a young firefighter with less than two years on the Department. Driving home that morning, I was looking forward to a nap to recoup from a busy night in the firehouse. When I arrived home around 8:00 a.m., the couch was calling my name. Just as I was starting to doze, the fire department dispatcher called.

Fire at the Mall

My division was being called to a multiple-alarm fire at a local mall. I quickly drove to my station to pick up my bunker gear, then downtown to station #1, where a lieutenant met me. He loaded his gear into my car, and we headed to the fire scene. About a half-mile away, when we drove through an intersection that was heavily charged with smoke, we started to realize the magnitude of the fire.

A Real Challenge

Upon our arrival, heavy smoke was pouring out of the first-floor doors of a large, two-story department store at one end of the mall. The second floor had no windows, so we couldn't tell if that floor was involved. As one would expect at 9:00 a.m. on a Sunday morning, only workers occupied the store. Dozens of employees had been busy stocking the shelves with merchandise in anticipation of the upcoming holiday shopping season. Scattered groups of these employees now were standing in the parking lot.

The lieutenant and I quickly made our way to the fire building, where a District Chief met us. "We have to get a line to the second floor." He instructed us to take a 2½-inch hoseline up a ladder to a section of roof where we could gain entry into the second floor. With the help of another firefighter, we worked our way quickly up a ladder to the roof.

> *We fought our way through dense smoke.*

Once on the roof, we were able to enter the second floor through a door to an employees' lounge. Inside, we fought our way through dense smoke, which afforded zero visibility. We tried to use light cords with 500-watt lights, and they didn't help at all. Blindly we crossed the employee lounge and passed through a doorway that led out into the main store.

Now we were in the lingerie section on the second floor of the department store. Another firefighter, Larry, and I slowly dragged the line forward into the darkness. Unable to locate the fire because of the thick smoke and the sheer size of the store, we paused from time to time to listen for the fire or look for its glow so we could aim our fire stream in the right direction. The heat was increasing and the smoke was getting thicker as we moved forward, so we knew we were on the right trail.

Crawling forward, we came across several mannequins that had fallen over. Each time we felt a torso or various body parts, we were fooled for a split second into thinking we had found a victim. Still unable to find the fire, we paused again, trying to get our bearings. Suddenly something hit us in the head, driving us to the floor. We found ourselves buried in ceiling tiles with flame rolling over our heads.

Lost in an Inferno

The entire ceiling had collapsed on top of us, releasing the heat that had been trapped above. Instantly, our ears began to burn from the tremendous heat. I turned to look or feel for my partner, but I couldn't find him. I yelled out, but there was no answer. I stayed as long as I could, but in the days before protective hoods, my ears were burning up, and I had to get out fast. I tried to follow the hoseline, but it was buried under the fallen ceiling and I couldn't find it.

I had to get out fast.

With my belly to the floor, I headed in the direction I thought would lead to safety. With zero visibility, I was becoming increasingly desperate. Finally I saw a light up ahead and I dived out of the door and into the bright sunlight on the roof. Thank God, when I looked up, there was Larry. He had found the same door mere seconds before me.

By now, the fire had intensified and we reported our lack of progress to the Chief. There was too much fire and not enough ventilation. The second floor had no windows, and the roof was concrete with no skylights.

There was too much fire and not enough ventilation.

The Chief called for a police bomb technician to try to blow a hole in the roof. The bomb tech set several charges, all to no avail. He had been able to blow only a few small holes about the size of a basketball in the thick concrete roof deck.

We made another attempt at fighting the second-floor fire. This time we began a slow interior attack, aided by an additional 2½-inch hoseline, which was brought up and placed in another opening to the second floor. Slowly we pushed the fire back into the southeast corner of the second floor, where we were able to bring it under control. About this time, we began getting reports of several employees unaccounted for.

The battle had lasted so long that it was now almost 4:00 p.m. We were all exhausted. Just then we discovered the first two victims. Two female employees had collapsed and died a few feet from the top of the east escalator. We found out later that one of them was a young mother with a six-month-old baby at home. Over the years, I have wished that I had contacted their families, to tell them that we treated them with respect and did our best to save them.

Pushed to the Brink

Continuing our search, we found a total of seven dead victims over the next hour. The fire had spread so fast that most of them died at their workstations. After locating the seven bodies, the medical examiner set up a viewing area where family members could come and identify their loved ones. My District Chief assigned me and another new firefighter the difficult task of carrying each victim down to the viewing area. Chiefs are known to sometimes give these difficult tasks to new firefighters, probably to see if we can handle it. If we can, we will make it as firefighters.

The medical examiner had gathered the family members, maybe thirty people, in a private area. As we carried the first victim, they were able to see the body when we came down the stairs. I will never forget the sounds they made as we came through the door—painful moans, groans, and cries. These sounds haunt me to this day. One by one, we carried each victim down. By the time we got the last victim out, it was about 9:00 p.m. and my shift was relieved of duty.

> *These sounds haunt me to this day.*

Physically and mentally exhausted, I drove home thinking about all I had done and seen that day. While I was taking a long, hot bath, everything came flooding back and I became violently sick to my stomach. I realized that we may put off those emotions temporarily, but we can't keep them in forever.

Dealing with the Pain

In retrospect, I wish my Department and I had realized the importance of Critical Incident Stress Debriefings. At the time, I didn't understand the need for CISD, but now I recognize how much a counseling session was needed. Incidents like this can haunt us for life. After twenty-six years, I hadn't been able to talk about this incident until now. A few years back, I started to tell my wife and it just wouldn't come out.

> *Everything came flooding back.*

About ten years ago I had to go back to the same department store to pick up concert tickets. When I found myself in the same area where we had found the first two victims, everything came flooding back. I had to get out of there.

Then, about a year ago, my son was shopping at that mall and his car broke down. He called me for help and I was standing there in the parking lot with my head under the hood of his car. Then I smelled the fire, and I was right back there on the second floor. I looked over and saw that the department store was being knocked down in preparation for a new store. The construction debris had retained some of that smell from twenty-five years before.

I also learned that a breakdown in accountability can be costly. When that ceiling came down on Larry and me, nobody knew where we were. We were lucky to have escaped with our lives. This was a breakdown in accountability and communication that most likely wouldn't happen today. It is up to all of us to ensure that it doesn't. Incidents such as this also reinforce the need for Rapid Intervention Teams. We must be proactive on safety.

This story puts an exclamation point on taking care of our own. Those who are making the decisions in the various Fire Departments across the country must realize that the most important thing they will ever do is to look after the well-being of their people.

Lessons Learned

- **Accountability is the key to safety on the fireground.** The Incident Commander must know where every fire company is working, and company officers must know where their firefighters are at all times.

- **Critical Incident Stress Debriefings (CISDs) should be utilized when needed.** We must look out for each other's physical and mental well-being. After extremely emotional or stressful situations, we should consider the use of CISD.

Discussion Questions

1. What tactical considerations could be used to increase the safety of the firefighters in a similar situation?

2. In today's fire service, what modern tools and equipment could be utilized at a similar incident?

3. How can we recognize the need for Critical Incident Stress Debriefings?

For Love of My Father

District Chief Robert Hoff
29 Years in the Fire Service

Chief Thomas Hoff

My twenty-nine years with the Fire Department have been like a roller coaster ride. The next incident is always just around the corner. Over the years, I've had a lot of experiences, from exhilarating to horrifying. Our teams have saved a lot of lives, and there is no greater feeling. We deal with the loss of life on a regular basis, and this is always hard to take. Also, because of the inherent dangers of our job, firefighters sometimes lose their lives in the line of duty, and this is the worst feeling imaginable.

Throughout my entire career I've worked in the busiest areas of the city, and every day I learned something new. The most powerful of the lessons I have learned happened more than fourteen years before I joined the Fire Department. It was the death in the line of duty of my father, Chief Thomas Hoff, on February 14, 1962.

The Saddest Day

I remember the day like it was yesterday. I was six years old and stayed home from school that day with the measles. My father and I sat at the kitchen table while he waited for his driver to pick him up. For the previous six months, my father had been the Assistant Drill Master at the Fire Academy. Prior to that, he was Chief of the 19th Battalion.

He was eating bananas and milk, and when his driver arrived, he pushed the bowl over to me and said, "Finish it for me, Bobby." I sat there eating what was left of his breakfast as I watched him walk out the door. That would be the last time I would ever see him. To this day, I have never eaten bananas and milk again.

A few hours later, the phone started ringing at our home. Anxious people who had heard news of a building collapse were asking about my father. "Is Tom at that fire?" My mom was answering call after call, "No, I don't think so." I could sense that she was getting nervous, probably hoping that the next ring would be her husband calling to say he was all right. That wasn't to be.

Missing

The doorbell rang, and I peeked around the corner as my mom ran for the door. It was my uncle, the lieutenant of Squad 8 at the time, and he had been at the fire. The Fire Commissioner had lent my uncle his car to pick up my mother and bring her to the fire. To this day I can still hear the pain in his voice. "There's been a building collapse, and he's missing," he said.

She quickly left with my uncle while I stood looking out the window and watching the shiny black car disappear down the snow-covered street.

> *The look on her face told a different story.*

My noticeably nervous aunt kept trying to assure me that they would find my dad and everything would be all right. The look on her face told a different story. The house quickly filled up with relatives. I remember the crying, praying, and a lot of whispering.

My father and another Chief had responded to a fire in a large four-story apartment building. It was frame construction with a brick veneer, more than a hundred years old. The building was totally involved and had been burning for quite a while.

In those days, company officers didn't carry radios. Sensing collapse, the two Chiefs entered the building to evacuate the firefighters. The last of the firefighters were exiting when the building collapsed and buried the two Chiefs and the Chief's driver. They had been in the kitchen walking

This photo shows the magnitude of the collapse. Photo courtesy of the Chicago Fire Department.

toward the rear door when it collapsed. The driver was buried not too far from my father. Buried alive, they couldn't see each other, but they talked for almost forty-five minutes. Then the building shifted and the driver didn't hear my father again. After a couple of hours, firefighters were able to rescue the driver.

At home, we sat around for hours, nervously waiting. Whenever the phone rang, everyone jumped. Each time, it was just another concerned friend offering support. My mother stood vigil at the fire scene for more than five hours while the search crew dug through the ruins, looking for her husband. They had to bring in a construction crane to lift the large sections of the building that had buried him.

Finally, after five and a half hours, they found his body. When the phone rang this time, it seemed to echo as loudly as a church bell in the quiet house. When my aunt grabbed the phone, all ears strained to hear the message. Her eyes welled up with tears, her lip started quivering, and we all knew.

I entered the Fire Department fourteen and a half years after the death of my father. Throughout my career, I crossed paths with a Chief who was at

A recovery team working near the spot where Chief Thomas Hoff was found. Photo courtesy of the Chicago Fire Department.

> *He told me it was time to talk.*

the fire when my dad died. He told me that someday he would sit down with me and tell me about my father. Finally, in 1997, on his last day on the job, I stopped by to wish him well and he told me it was time to talk.

A Recounting

At the time of the fire, this man was a lieutenant on a squad company. He and another firefighter were working on the fourth floor with a few other companies. Heavy fire was in the walls and rolling in the cockloft area. They were ordered out of the building, but he yelled out of a window, "Just give us five more minutes—we're getting it."

The Fire Commissioner, Robert J. Quinn, told my dad and Chief O'Brien, "Get them out of there."

The two Chiefs followed the hoselines into the smoke-filled building and started ordering the firefighters out of the building. Lowering his head, he told me, "Your dad and Chief O'Brien saved my life." He and his squad company were the last firefighters out the door, he said. They had just stepped onto the back porch when the roof came down.

He told me, "Bobby, all these years I've lived with this." Our eyes both welled up as he went on, "We thought we could get it—just five more minutes."

As he finished his story, I felt badly for this man. He had lived for thirty-five years with guilt. I told him that I respected him and didn't hold him responsible. He was just being an aggressive firefighter, and I've made that same mistake myself over the years.

It reminded me of an incident several years ago when I almost caused some firefighters' deaths by my being overly aggressive. I was the lieutenant of Squad 2, and there was a fire in the basement of a grocery store. A chief ordered us out of the building, and I responded over the radio, "Let us take a shot at this, Chief . . . just five more minutes."

He said, "No—bring them out now." We came out, and five minutes later the terrazzo floor collapsed into the burning basement. This is a lesson I should have learned from my father's story, and it was reinforced years later: When someone orders us out of a building, we must get out.

I hope my father's death will serve as a warning message so others may live.

Lessons Learned

- **In risky situations, we must think of our families.** When a firefighter dies, many people suffer. Family, co-workers, and friends all have to deal with the loss of life. We must keep this in mind when we are tempted to take needless risks. Risk management is essential.

- **We shouldn't second-guess any order that deals with firefighters' safety.** No building is worth the life of a firefighter.

- **There is no greater sacrifice than to give up one's life for another.** Chief Thomas Hoff and Chief Robert O'Brien gave their lives in the line of duty on February 14, 1962. They risked their lives to save many lives that day. God bless these heroic men. May they always be remembered.

Discussion Questions

1. What type of emergency evacuation order does your Fire Department in your area utilize?

2. How does good communication tie in with accountability?

3. If a similar incident were to happen today, what modern-day equipment and procedures might assist in saving lives?

⌒ Final Thoughts ⌒

To the everyday citizen, I hope these stories enabled you to see through the smoke and look into the hearts and minds of firefighters. The average firefighter doesn't brave the everyday dangers looking for an award, or to be honored in a tickertape parade. The everyday firefighter is content with the simple rewards in life—a pat on the back from their peers for a job well done, or a warm cup of coffee on a cold night. These everyday heroes appreciate how precious life is because they see so much death. After each tour of duty, they go home to their families knowing in their hearts that they are making the world a better place, one day at a time. This is their reward.

To my brother and sister firefighters, please use this information as an educational tool. Many of the lessons in this book came at a high cost, in some cases by firefighters making the supreme sacrifice. Never forget these heroes or the lessons we learned from them. We have learned how to increase safety on the fireground to protect lives in the future.

Please e-mail me with any questions or comments. Let me know if this information helps you at an incident, or if you want to share your experiences for future projects. I welcome all comments, and will reply to each.

Thank you and stay safe,

Steve Chikerotis
e-mail: FFHeart343@aol.com